The Nature of Ball Lightning

Stanley Singer
Director, Athenex Research Associates
Pasadena, California

℗ **PLENUM PRESS** • **NEW YORK–LONDON** • **1971**

Library of Congress Catalog Card Number 70-128512
SBN 306-30494-5

© 1971 Plenum Press, New York
A Division of Plenum Publishing Corporation
227 West 17th Street, New York, N.Y. 10011

United Kingdom edition published by Plenum Press, London
A Division of Plenum Publishing Company, Ltd.
Davis House (4th Floor), 8 Scrubs Lane, Harlesden, NW10 6SE, England

All rights reserved

No part of this publication may be reproduced in any form
without written permission from the publisher

Printed in the United States of America

Preface

In 1837 a comprehensive discussion of lightning appeared in the *Annual* of the French Bureau des Longitudes with a section on ball lightning which provided for the first time a readily available source in the scientific literature of the basic properties of this curious natural phenomenon. The author, Francois Arago, was the dominant influence in the French Academy of Sciences in the nineteenth century, having become a member of that august body at the age of twenty-three. His attention alone doubtless served at that time to establish the validity of scientific interest in the problem. In addition his discussion covered some of the major questions associated with ball lightning in a notably clear-sighted, effective style. Later reconsideration of the same questions often provided no significant improvement over Arago's discussion.

There followed a dauntless band of varying but always small number who attempted to account for an apparently simple natural occurrence, a ball of fire usually seen in thunderstorms, with the best knowledge that advancing science could provide. All attempts to deal with this phenomenon were invariably frustrated. The characteristics of ball lightning could be readily cataloged, but they firmly resisted both experimental reproduction and theoretical explanation. One may say that to this day there is no explanation accepted by a large number of scientists. Several investigators of great ability and considerable accomplishment in different fields of science, including Faraday, Kelvin, and Arrhenius, took note of the problem. Arrhenius discussed ball lightning at some length in his textbook on cosmic physics; and the work which is now the standard, Brand's monograph of 1923, was published as a volume in a series on problems of cosmic physics. Ball lightning has been frequently characterized by recent investigators as a plasma sphere. The brief but influential paper by Kapitsa was responsible for arousing my own interest, which was soon directed to the question of the structure of a plasma sphere of small dimension, a problem which still engages the efforts of my colleague Nak Goo Kim.

The persistent interest in ball lightning may be traced in the published works dealing with this difficult phenomenon over a period of one and one-quarter centuries. Recent publications reveal in general a knowledge of only a small portion of this information, and frequent inquiries indicate the difficult accessibility of Brand's monograph, which is now over forty-five years old. Many valuable observations and well founded conclusions from the past are evidently not readily available to modern investigators. One of the goals of the present work is the presentation and correlation of all the published information on ball lightning. The number of papers has varied sporadically in the past, periods in which great activity was exhibited alternating with sometimes lengthy intervals in which no papers were published at all. A marked increase in interest is evident at present. The literature has been covered through 1969, several investigators having kindly provided advance information on material which will soon be published.

An attempt has been made to describe ball lightning theories in somewhat the terms which made the theories seem of merit when they were first set forth. This should not be mistaken for undue sympathy for such hypotheses. Several, indeed, had little merit to begin with. The incredible variety of theories which have been proposed may cause some difficulty if not confusion. Perhaps, on the other hand, this approach in the discussion may reduce the frequency with which theories duplicating earlier, discredited explanations are proposed in the future; and some assistance may be provided for recognition of a theory which finally succeeds in accounting for all the properties of ball lightning.

Several specialized areas in meteorology, physics, and chemistry have been involved in consideration of the problem. The continuing lack of a conclusive theory occasionally prompted early if not actually hasty application of notable advances in these fields to the theoretical models of ball lightning. The discussion of such material (for example, from plasma physics, which has been given increasing consideration) presents an exercise in rationality. An attempt has been made to emphasize the physical meaning of results. The mathematical expressions, although more concise, are usually omitted in the discussion, which should not prevent their regeneration by the reader when desired. Here the goal has been to present the material from each of the fields involved in a readily metabolized form for those who may not be particularly attuned to a specific scientific area, as well as to the reader with general interests in natural science who may be attracted to this extremely curious and interesting phenomenon.

Scientific endeavors to solve the question of ball lightning are in marked contrast with several of the more extensively developed and formalized fields of study. Few investigators have been free of more pressing secular requirements to devote themselves to the problem, and continued study has certainly been discouraged by the difficulty it presents. The fraternity of scientists who have given some thought to the problem is, however, especially illuminated by the scientific traditions of liberal instruction, exchange of information, and vigorous discussion. The author has benefited greatly by this spirit and acknowledges the valuable assistance given by a number of colleagues.

J. Dale Barry of the Institute of Geophysics and Planetary Physics, University of California at Los Angeles, William J. Borucki of the Ames Research Center, and James Powell of Brookhaven National Laboratory reviewed the manuscript and suggested improvements. The comments of the latter were kindly provided by Gordon and Breach, Inc. Dr. Barry gave valuable aid in other aspects of preparation of the manuscript. He has established a comprehensive reference index for ball lightning maintained on computer cards with continuing additions from current publications.

Rida S. Bross, State University of New York at Buffalo, uncovered older references to ball lightning in the literature, and Y. Ksander, Library of Congress, assisted with many current papers as they appeared in the active foreign literature.

The courtesy of the following for permission to use photographic material and in some cases for providing difficult-to-obtain prints is gratefully acknowledged: The Burndy Library for the engraving of Professor G. W. Richmann's death; Sigurgeir Jónasson for the picture of lightning in a volcanic cloud; L. E. Salanave, Institute of Atmospheric Physics, University of Arizona, for the photograph of storm lightning; the American Association for the Advancement of Science for permission to use these two photographs which appeared as cover illustrations on *Science*, Vol. 148, 28 May 1965, and Vol. 151, 28 January 1966 (Copyright 1965 and 1966 by the American Association for the Advancement of Science); E. J. Workman and R. E. Holzer for the photograph of a lightning discharge near the ground; the U.S. Naval Ordnance Laboratory for the film of lightning striking an underwater explosion plume; B. T. Matthias and *Nature* for the photograph of pinched lightning; *Umschau in Wissenschaft und Technik* for the photograph of street lights made with a moving camera; L. A. Vuorela and *Geophysica* for the photograph of ball lightning in motion; and E. Kuhn and D. Kuhn and *Die Naturwissenschaften* for the photograph of ball lightning emitting sparks. David L. Singer assisted with the figures.

The steadfast encouragement of James Hughes played a large role in both the initiation and the continuing stages of the study which led to this volume. A major portion of the work was supported by the Office of Naval Research.

Contents

Chapter 1.	Storm Lightning and the Problem of Ball Lightning	1
Chapter 2.	Ball Lightning in the Prescientific Era	5
Chapter 3.	The Properties and Processes of Storm Lightning	11
Chapter 4.	The Question of the Existence of Ball Lightning	18
Chapter 5.	Observations of Ball Lightning	23
A.	Spherical Shapes Associated with Lightning Flashes	23
B.	Individual Accounts of Ball Lightning	27
C.	Collections and Reviews of Ball Lightning Observations	48
Chapter 6.	Photographs of Ball Lightning	51
Chapter 7.	Characteristics of Ball Lightning Derived from Observations	62
Chapter 8.	Theories and Experiments on Ball Lightning	77
A.	Agglomeration Theories	78
B.	Leyden Jar Structures	78
C.	Transformation of Linear Lightning into Ball Lightning	80
D.	Generation of Ball Lightning by Chemical Reactions	81
E.	Nuclear Theories	88
F.	Charged Dust and Droplet Models	89
G.	Molecular Ion Clouds	92
H.	Vortex Structures	94
I.	Ball Lightning as an Electrical Discharge	98
J.	Luminous Spheres from Vaporized Solids	111
K.	Plasma Theories and Experiments Applicable to the Problem of Plasmoids	114
L.	Plasma Models of Ball Lightning	125
M.	Formation of Ball Lightning by Natural Electromagnetic Radiation	133
Chapter 9.	Present Aspect of the Ball Lightning Problem	146
References		149
Subject Classification of References		165
Index		167

Chapter 1

Storm Lightning and the Problem of Ball Lightning

The immense thunderstorms which occur in nature contain huge electrical machines whose activity is displayed in lightning discharges. The intensity of electrical activity is beyond question to observers who are left with ears ringing and eyes blinded, such as the witness to a cannonade of flashes who noted two hundred lightning strokes in a minute or the one who reported displacement of a massive stone wall over several feet by a single discharge.

In ancient times lightning was viewed with well-deserved trepidation. The Etruscans reputedly called down bolts from the skies on opposing armies. Several forms of lightning were recognized: linear, zigzag, sheet lightning, and ball lightning. Benjamin Franklin in a superb exercise of scientific insight recognized the similarity between the effects produced by the eighteenth century electrical friction machines and ordinary lightning. The simple, direct experiment he suggested to test his theory, that of bringing down the electrical charge from the skies to easy access by the investigator by means of a conducting line carried by a kite, soon established that lightning is an electrical discharge.

A century later C. V. Boys conceived a rotating lens camera which provided a method of discerning the rapid sequence of events leading to a lightning stroke. A questing, almost invisible electric probe leaves the highly charged storm cloud traveling in rapid, short runs with intervening pauses, seeking a favorable path to the ground or to another cloud. With a channel established, the intense major discharge follows, releasing currents as great as one hundred thousand amperes at a billion volts potential between the cloud and ground in a small fraction of a second. This discharge produces the bright flash seen as the lightning stroke.

Only recently have cloud processes of charge separation been recognized which seem capable of generating electricity in the magnitude evidently involved in lightning. The rate of charge production estimated from any single process is, however, barely adequate to account for even moderate rates of lightning discharge which have been observed. Charge separation is attributed to collision of raindrops, freezing, friction of drops or solid particles with air, breakup of drops on collision, and similar gas–particle interactions. Although reasonable charging processes are known, the actual cloud processes have not been established.

The ordinary lightning discharge is fairly well understood in principle, and considerable progress has been made in determining the detailed mechanism by which it occurs. This is far from the case with ball lightning, as far as the world of science is concerned. Although the phenomenon has engaged the attention of leading scientists such as Boyle, Arago, Faraday, Planté, Lodge, Arrhenius, Toepler, and Kapitsa, little unanimity of opinion is found on this seemingly analogous thunderstorm occurrence. The reasons for the controversial state of knowledge of ball lightning are readily evident from study of the information available.

Reports of ball lightning are found in ancient literature as far back as written records mentioning weather phenomena exist and, before that, in ancient art. Many modern investigators of the problem are unfamiliar with significant observations made even one hundred years ago. Discussions of the problem in the last few years, purporting to be complete, are based on a relatively small portion of the available information and reviews largely published in the last twenty-five years; whereas an evaluation of the problem which has been unsurpassed for its critical judgment and detail appeared over fifty years ago. This may account in part for the frequent presentation as new of theories which were actually suggested some time in the past. Wholly incorrect statements have been made in otherwise authoritative discussions, even when reliable data were available in the contemporary literature. Few problems of nature have shown as difficult progress as this one with the advance in general scientific knowledge.

Ball lightning is a luminous globe which occurs in the course of a thunderstorm. It is most often red, although varying colors including yellow, white, blue, and green have also been often reported for the glowing ball. The size varies widely, but a diameter of one-half foot is common. Its appearance is in striking contrast to ordinary lightning, for it often moves in a horizontal path near the earth at a low velocity. It may remain stationary momentarily or change course while in motion. Unlike the rapid flash of ordinary lightning, ball lightning exists for extended periods of time, several seconds or even minutes. Its path in many cases takes it indoors where the glowing ball occasionally passes quite close to an observer. It enters by a window or a chimney and may depart through such openings. Prof. Colin Eaborn of the School of Molecular Sciences, Sussex University, remembers the windows of his childhood home being left open during thunderstorms so that ball lightning, if it appeared, could leave without difficulty. Zbigniew Zielkiewicz recalls that as a boy the windows of his home were closed during storms to prevent drafts from drawing the fireballs inside. In many cases witnesses note particularly that, although intensely bright, the ball gives off no heat and disappears silently. In others a great explosion occurs, displacing and damaging nearby objects.

Marked variation in these general characteristics is noted. The ball is seldom a true sphere; often it is a mass with only a generally defined form or with some protrusions. Sparks may be emitted by the body. The boundaries are sometimes well-defined, sometimes hazy and shrouded in mist. A hissing or crackling sound as in an electrical discharge is often reported, or the ball may be silent. It falls directly from cloud to ground like a body of some mass, hovers, or even rebounds from the earth like an elastic body. In some cases it seems to be borne by the wind; in others it travels directly opposite the prevailing wind direction.

The great variation in properties reported has been a source of much confusion in providing a clear explanation of ball lightning. Theories have been unusually numerous. Most explanations accept the role of storm electricity at least in initiating the glowing mass. The continuing activity of the ball may then be ascribed primarily to chemical reaction or to electrodynamic processes. Chemical theories presented historically include suggestions that nitrogen triiodide, hydrogen and oxygen mixtures, or ozone formed by storm discharges provide the substance of the ball and the energy released in subsequent decomposition. Active nitrogen produced by a lightning stroke and a nitrogen discharge "flame" in the atmosphere resulting in nitrogen oxides have been proposed. Wholly electrical theories have presented the ball as a brush discharge. The separation of a short segment of an ordinary stroke of lightning as a vortex has been suggested. Vaporization of a metal such as copper by an intense lightning stroke may produce the ball. Electrically charged dust particles, raindrops, or ions of the atmospheric gases in assemblies in which neutralization of opposite charges by recombination is somehow slowed have been discussed. Plasma theories provide many of the most recent models considered, based on the rapidly increasing knowledge of matter at high temperatures from thermonuclear research.

Since each theory succeeds in explaining at least a few properties of the ball often ignoring others, conclusive rejection of inadequate theories on definitive grounds is seldom found. Occasionally widely different explanations have been presented by a single author with no comment on the error of his previous model and no comparative discussion showing an advantage in the newer one. To an unusual extent new theories have been offered with insufficient consideration of previous work, of simple physical laws, and of the information available describing this phenomenon.

In order to deal with this problem in a systematic way, the theories of ball lightning will be considered in the following discussion. These can be assessed on the basis of the now considerable information which has accumulated on the many new questions encountered. At the same time, in the course of this inquiry some simple and fundamental questions were encountered in several areas of science where information is lacking, and the desirability of future investigations of general interest has become apparent. Specific grounds can be cited for the rejection of almost all of the historical theories with the information now available. Most unexpectedly, study of numerous sightings (well over one thousand are represented in the literature) and consideration of some laboratory experiments permit a definite conclusion from data which have been available for over fifty years on the identity of one form of ball lightning. Very recent results have provided valuable information on a relatively new theory which shows promise of improving our understanding of this phenomenon.

Chapter 2

Ball Lightning in the Prescientific Era

Lightning flashes appear in divers forms during thunderstorms. The observation of the different shapes led to their classification as different types of lightning even in ancient works which no longer exist and are only known from references in later writings. Three general forms, for example, are represented by the ordinary lightning stroke, often called zigzag lightning, which is seen in a bright well-defined path; sheet lightning, which is faintly luminous and covers a broad region of the sky; and ball lightning, which glows either faintly or very brightly but presents a very distinctive appearance both in shape and in its varied paths. Etruscan art shows such types of lightning, including the ball of fire.[160] A discussion of ball lightning is attributed to Aristotle in the fourth century B.C. in his work on meteorology.[17]

Aristotle's references to a slow-moving thunderbolt and to lightning formed in enclosed spaces are not clearly concerned with ball lightning. Seneca refers to the work of Posidonius in the first century B.C., who noted six classes of lightning including ball lightning.[176] Lucretius has been credited[102] with an early discussion of ball lightning in his work on natural phenomena,[294] but here too some ambiguity is evident. Lucretius states that ordinary lightning is composed of hard, spherical, bright atoms which enable it to penetrate hard materials.

The fiery flight of high-velocity bodies such as meteors or bolides which enter the earth's atmosphere from space was often confused with electrical phenomena arising in storms. This difficulty has been noted in recent meteorological compilations[462] as well as in historic works.[240] The distinction in general properties between a solid body flying through the skies and a glowing ball arising as some manifestation of an electrical storm is, in general, easily made. The solid masses traveling in the atmosphere at high velocity fly in straight paths. Near the end of the flight when their mass becomes greatly reduced the path may suddenly show repeated oscillations. A characteristic luminous trail made up of material eroding from the main mass extends for some distance behind it. Portions of the initial body may separate and travel off on divergent paths. The whole mass may be eroded away before it can strike the ground, or it may become molten and strike when it is almost entirely in the liquid state. A solid portion of these bodies often reaches the ground. The flight of meteors through the atmosphere occurs in a storm only by coincidence. Many of these characteristics are in marked contrast to those we have noted for ball lightning. Ball lightning is usually associated with storms, although as with other lightning forms (e.g., heat lightning) it may be observed in the absence of storm conditions. While we have not established by any direct method the matter of which the ball is composed, its formation by some action of natural electricity and the complex paths often displayed near the ground at low velocity indicate that a solid core is unlikely.

Although the term "electrical meteor" was often used for ball lightning, a distinction between this and other luminous objects in the atmosphere was made early by Muschenbroek,[352] who is credited with the invention of the Leyden jar. Confusion of meteors with ball lightning continued, however, as indicated in observations erroneously reported as ball lightning as much as one hundred fifty years later. Occasionally ball lightning, which is relatively rare, has been reported as a meteor.[65, 240] The continual reappearance of the problem of correct identification of fireballs as meteors or ball lightning has resulted from many reports in which the observed properties of the ball were not clearly those of one phenomenon or the other. For example, the flight of slow bolides (or "Bradytes") was reported several times late in the nineteenth century.[83] These objects were evidently meteors with the exception that they traveled at very low velocity. One case was apparently observed in widely separated locations by two witnesses, one of whom had seen ordinary meteors. Since all other aspects of these occurrences, few as they may be, are in accord with their identification as meteors, this classification seems most acceptable; the exceptional speed and long duration of these specific cases are explained, as usual in such problems, as due to unusual trajectories and the accompanying difficulty of estimating velocity under such conditions. For example, a high altitude path nearly parallel to the surface of the earth might account for the reported observations.

Early observations of ball lightning provide very limited historical material. On the other hand, the modern era in the study of this problem, using the scientific method which we recognize today, begins early considering the difficult obstacles to systematic study in this field. Three sources of information are of value: observations of ball lightning in nature provide the basic material, rationalizations of the observations or theories attempt to compare the properties of ball lightning with other objects; and experimental studies in the laboratory, primarily with different types of electrical discharges, attempt to duplicate the properties of ball lightning, often according to the conditions suggested in the theories. One may say at the outset that no experiment planned for the purpose has succeeded in duplicating all of the properties of the natural object.

The characteristic properties of ball lightning will be presented later by means of selected observations which have been collected from the literature. A few of the earliest cases which are primarily of historical interest are given here. The earliest description[62] of what may be an appearance of ball lightning in the sixth century A.D. was found in the works of St. Gregory of Tours. A fireball of blinding brightness appeared over a procession of religious and civil dignitaries of Tours during the dedication of a chapel. The sight was so terrifying that all the people in the procession threw themselves to the ground. Gregory's discussion of this event indicated that he was acquainted with similar occurrences. He mentioned lights emanating from the hair of people in certain localities and seasons of the year which would now be recognized as St. Elmo's fire, the discharge from an object in a high electric field. Gregory also discussed imaginary radiation and flames seen from religious relics by few people in the presence of many others for whom such lights were invisible. He noted that the fireball seen over the procession was visible to all. Since there was no reasonable explanation for the ball, Gregory concluded it was a miracle.

An entry in the Anglo-Saxon Chronicle for the first week of the year 793 A.D. describes a heavy storm in which "fire dragons" were observed in additional to intense lightning, the latter presumably the usual zigzag discharges.[12]

In memoirs from the period[41] a flame was reported entering the bedchamber of Diane of France by the window on the night of her wedding, March 3, 1557. The light passed around the room and finally came to the bed, where it burned her hair and bedclothes.

The stationary luminous globes of St. Elmo's fire were often seen fixed to the tips of the masts of sailing ships. Several observations were also made of of ball lightning at sea which traveled considerable distances. A blue ball of fire the size of a large millstone was observed three miles from the *Lizard* in 1749 rapidly rolling down on the ship.[96] It came within fifty yards before they could raise additional sail, rose almost straight up, and exploded with the sound of hundreds of cannon. The smell of brimstone was very strong. The main topmast was shattered, and the main mast was split to the heel. Spikes in the main mast were drawn out and driven into the main deck with such force that the carpenter had to use a crowbar to remove them. Five men were knocked down and one burnt badly by the explosion.

The early communications on Franklin's hypotheses concerning the electrical nature of lightning inspired widespread studies on the electrical activity of storms by a new group of natural philosophers, who came to be

Fig. 1. The death of Richmann. [Burndy Library Collection]

known as the Electricians. The death of Professor G. W. Richmann in the summer of 1753 in St. Petersburg has been ascribed to ball lightning, or in a recent Russian description of the event,[102] to the first experiment which succeeded in producing artificial ball lightning. The "experiment" was witnessed by a friend of Professor Richmann, the engraver to the Royal Academy of St. Petersburg. In addition to the engraving made by this direct witness there is available a description communicated to the Royal Society of London containing information contributed by Lomonosov, who visited the laboratory and made a detailed inquiry.[561] Richmann was observing with his friend, the engraver, the effect of a storm on his device for measuring the electrical field of the atmosphere. Witnesses outside the laboratory saw lightning hit the metal rod on the roof which was connected to the measuring apparatus located in Richmann's laboratory. Inside, a ball of blue fire the size of a fist came from a metal rod on the apparatus straight to Richmann's forehead as he stood approximately one foot away (see Fig. 1). There was a report as loud as a pistol shot when the globe hit Richmann, and hot wires from the apparatus struck and burned the other's clothes.

Despite the frequently observed association between ordinary lightning discharges and ball lightning the earliest theories posed by the Electricians were not electrical theories. Muschenbroek[351] suggested that ball lightning was a collection of inflammable gases. Faraday[147] noted that the velocity and duration of ball lightning was incompatible with all the known properties of ordinary electrical discharges, concluding that any relationship to lightning or atmospheric electricity is "more than doubtful." On the other hand, the formation of a small rotating red ball of fire in an experiment with the Leyden jar was presented in the late eighteenth century[46] as an explanation of bolides, which we now know to be solid bodies heated to incandescence by their rapid flight through the atmosphere. The fireball in the Leyden jar ended with a loud explosion, and a hole was cut clean in the side of the flask. A strong sulfurous odor was noted afterwards.

Arago,[16] who was elected to the French Academy at the age of twenty-three, gave the first comprehensive and rational discussion of ball lightning in the course of a paper on lightning in the annual publication of the French Bureau des Longitudes in 1838. Arago commented years later that he had been cautioned against entering into this field which had been so thoroughly covered by Franklin. Arago's work, entirely in the spirit of modern reviews of this problem, included a general description of ball lightning characteristics, a collection of more than twenty reports of ball lightning observations, augmented in later versions by selected examples of numerous additional observations which he received after his interest in this phenomenon became known, and a suggestion as to the possible identity of the balls. He enumerated some of the difficult questions which science had not yet answered and considered the question whether ball lightning really exists or is an optical illusion. His affirmative view of the reality of ball lightning and his theory that it is partially the substance formed in lightning will be discussed in later sections dealing with these questions. In a volume on thunderstorms published five years after Arago's work, Snow Harris[200] suggested that ball lightning is an electrical brush or glow discharge. With Arago, however, the problem of ball lightning entered the province of science.

Chapter 3

The Properties and Processes of Storm Lightning

Many observations of ball lightning indicate a close association with ordinary lightning strokes in the thunderstorm. Although the earliest theories were not electrical even after the role of electricity in ordinary lightning was well known, the close association of ball lightning with storms in which electrical activity was especially marked soon led to the idea that the different forms of lightning were closely related. Thus, ball lightning was described as stationary lightning,[240] and ordinary lightning was said to be the trajectory of rapidly-moving ball lightning.[348, 494] The references in Aristotle to slow-moving thunderbolts in contrast to fast-moving ones and in Lucretius to lightning vortexes may indicate that they also considered these as different forms of the same substance in nature, not as different storm phenomena. The frequent resemblance of ball lightning theories through the years to views held on common lightning discharges, including the processes and substances which play a major role, encourages consideration of the comparatively well understood linear form.

The discharge of storm lightning, a continuing renewal of Prometheus' gift of fire to man, occurs as great potentials built up by charge separation processes in clouds cause the sudden flow of intense currents by which the potentials are dissipated. The intense light of the stroke is radiated from the path in which the current flows, defining a region which is distinct from the surrounding atmosphere (Fig. 2).

The electric field may differ between a storm cloud and the ground by as much as 10^9 V, as the electrical field gradient in fair weather, which may be 100–400 V/m (less over the ocean, greater over uniform flat land areas), increases to many kilovolts per meter in stormy weather.[66, 97, 461] The use of these high storm potentials as a natural laboratory for electrical experiments has been suggested.[66] The average current flow in the stroke is of the order of 10,000 A although the largest peak currents observed[97] have been

Fig. 2. Storm discharges. [*Science* **151**, No. 3709 (1966).]

over 100,000 A. The total quantity of charge has been reported[97, 214] from 0.02 up to more than 100 coulombs. Twenty coulombs is an accepted average value. The electrical power flowing in the lightning channel according to these parameters may be as great as 10^{12} watts and the corresponding work approximately 10^9 joules.

High-speed photographic and electric observations have shown that there are several detailed processes contributing to the gross electrical parameters given above in each lightning flash. After the formation of a given lightning channel, often three or sometimes over forty strokes may flow in the same path. The combined time for the discharge of all these pulses totals approximately 0.25 sec, with an interval of 10–100 msec between pulses. The lightning flashes made up of multiple strokes can show long periods of luminosity from 40 msec up to a reported 0.27 sec duration.[253] The long-lasting glows occurred in channels which maintained sufficient conductivity to allow an increase in current without repetition of the leader process and were thus associated with continuing current. Radiation from the molecular nitrogen ion $N_2{}^+$ was observed with a photometer for up to 0.8 sec after lightning.[344]

The 3914-Å band ($B^2\Sigma u^+ \to x^2\Sigma g^+$) is the one involved; this ultraviolet wavelength is not visible. The greatest intensity was approximately 50 rayleighs (5×10^7 photons/cm^2 sec) above the airglow background. The long-lived radiation was ascribed to molecular and resonance scattering of the 3914-Å light from the lightning, however, rather than direct light from the flash itself.

In discharge of ordinary zigzag lightning the channel is formed by an initial flow of electron charges or leader which leaves a cloud with a velocity of approximately 10^7 cm/sec running in steps perhaps 50 m long with 50 μ sec intervals and, thus, by a rather wandering path finds its way down close to the ground where the electric field from the leader becomes large enough to cause electrical breakdown of the air, completing a channel of sufficient conductivity for the ensuing large electrical pulses. The lightning path from cloud to ground often covers a distance of 5 km. When formation of the channel by the leader is complete, a sudden large pulse of current from the earth proceeds up the channel at up to 10^{10} cm/sec, producing the intense light flash which we see. The short duration of this return stroke, 100 to 115 μ sec, limits the total amount of charge which is actually transferred in this portion of the process despite the high peak currents.[365] Thus, the first leader may transfer twice as much charge as the return strokes. Up to 20 coulombs may be carried by each.[534] The electrical breakdown potential to give small sparks in air containing water droplets has been estimated at 10^6 V/m, but the natural cloud process involving the electron leader which is emitted to establish a discharge channel may occur at lower electric-field gradients.

Lightning flashes which differ from the general processes just described are also well known. Not all flashes are preceded by a leader. Flashes transferring electrons to the ground are those occurring with multiple strokes, whereas positive flashes most often have only a single stroke.[44]

The properties of the return stroke of lightning which produces the intense light in the discharge channel are of special interest since ball lightning is often considered to be the same material in a different structure. The diameter of lightning has been measured by the size of holes melted in fiber-glass screens.[530] The perforations fell into two general classes, one relatively small (2–5 mm in diameter), the other ten times as large (2–3.5 cm in diameter); but it appears that the screen holder may attract nontypical discharges of lightning.[290] Measurements from photographs[144] made at distance of approximately 100 m gave diameters of 3–11 cm.

A peak temperature in the lightning channel of 24,000°K was measured from the optical spectrum of a return stroke by the relative line intensities of nitrogen and oxygen atoms and singly ionized nitrogen.[531] Time resolution of spectra[366] from the return stroke to 5 μsec intervals indicated a maximum temperature of 30,000°K decreasing to 16,000°K in 30 μsec. The mass density of the channel was calculated[531] as one-tenth that of air at standard conditions, but the pressure of the neutral and ionic particles was approximately 18 atm. The time-resolved spectra[366] gave a peak pressure of 8 atm in the first 5 μsec, decreasing to 1 atm after 20 μsec. Ionization was nearly complete in the first 15 μsec, decreasing thereafter. In the values derived from determination of the peak conditions there were approximately 4×10^{18} electrons/cm^3 in the discharge channel (approximately 0.81 electron for each air particle of all other types, either neutral or ionic), and the pressure in the stroke caused by the electrons alone was 14 atm. The singly ionized atoms of nitrogen and oxygen, N^+ and O^+, were present as the major ionic species at the ratio of 0.64 and 0.16 for each air molecule. The molecular nitrogen ion N_2^+ at 7.2×10^{-6} and NO^+ at 5.8×10^{-6} relative concentrations were the other major ionic species observed.[531] The mass density and pressure parameters were derived from the optical spectrum, and the stroke temperature was derived from this spectrum and the thermodynamic properties of air. Several assumptions are involved in determining the parameters in this way, but the values obtained are in good agreement with previous careful estimates. The visible light radiated from the lightning flash is primarily given off by neutral nitrogen and oxygen atoms and the monopositive nitrogen and oxygen ions which are the major ionic species. Some neutral and ionic molecular nitrogen states contribute to the luminosity, as well as CN, hydrogen α, and argon spectra.[534]

The mechanisms by which the storm cloud generates the high electric fields and currents causing the intense discharge just described are still a difficult problem in the study of atmospheric electricity on which there is little accord. All theories consider charge separation processes for the production of the required charged current carriers from atmospheric gases, rain droplets, and ice particles, which are initially neutral. The great magnitude of the charge flow in the numerous lightning discharges in a storm presents a special difficulty. Over 200 flashes/min have been reported[51] in some storms. Several reasonable charge separation mechanisms have been suggested, and there is considerable experimental evidence showing that these processes do indeed give separation of charge.[108, 307] The collision of water drops of different size in a low electric field such as that present even in fair weather induces charges on the drops. Drops falling in such a field can also capture free ions or electrons from the atmosphere. Collision of drops in an electric field can result in breakup with accompanying charge separation.[189] Similar processes have been considered with ice particles in the cloud. The freezing of water in clouds occurs after supercooling as indicated by related properties such as the temperature and electric field profiles with altitude; supercooled water exists down to $-40°$C. Ice particles can undergo the processes attributed previously to raindrops. In addition, experiments have shown that freezing itself causes charge separation.[585] The coalescence of supercooled raindrops with particles of hail followed by freezing and splintering has been suggested as the major process in thunderstorm electrification.[108, 307] A thermoelectric effect is displayed by ice in which a temperature gradient produces separation of charge.

Despite the goal of most investigators to establish one of these charging mechanisms as the predominant one occurring in nature, no single process is sufficient to account for the quantity of current involved in active thunderstorms with high lightning frequencies. Extremely high rates of charge separation are indicated both in the appearance of lightning discharges within a few seconds of the beginning of the storm and in the high rate of occurrence of discharges. Some of the processes given are definitely more effective than others under suitable conditions, for example, the collision of supercooled water drops with hail particles can provide charges sufficient for a moderate lightning rate. Warm clouds in which freezing cannot occur because of the

Fig. 3. Volcanic lightning. [S. Jónasson, *Science* **148**, No.3674 (1965).]

temperature, however, also display marked lightning activity.[108] Such difficulties may necessarily lead to an increasing consideration of charge separation by several processes operative concurrently in the cloud structure. The further question of how charge once separated by these mechanisms can be transported to relatively distant regions of the cloud has led to theories which emphasize the importance of circulation or convective flow in clouds.[553, 585]

All the mechanisms for charge separation are evidently heterogeneous processes involving particles, either raindrops or ice, snow, hail, etc. The close relationship of cloud electrification and such particles in general may be indicated by observations made over a century apart of rain gushes following lightning strokes in certain clouds.[103, 343] The occurrence of bright electrical discharges also designated as lightning in volcanic clouds (Fig. 3) indicates that charge separation is effective with solid particles in atmospheric

Fig. 4. Complex lightning path near the ground photographed with moving film. [R.E. Holzer and E.J. Workman, *J. Appl. Phys.* **10**, 659 (1939).]

processes possibly including gas–solid and solid–solid collisions, even in the absence of liquid or liquid-freezing processes[11]. Two methods of formation of positive charge were suggested by the investigators of volcanic eruptions, production of salt particles in the air by contact of molten lava with sea with the walls of water, which has some basis in laboratory results, and friction of the particles the crater throat as they escape. Electrification from land volcanoes indicates that the role of sea water is not essential. The separation of charge on micron-size solid particles blown by gases has been demonstrated experimentally.[264, 289]

Particle processes have thus been given a major role in theories on how storm clouds produce the electrical charges required for lightning activity.

Different discharges of lightning vary greatly, not only in appearance but also in the electrical parameters discussed above for which, in general, representative maximum values or mean values have been given. Lightning discharges in some regions of the world display specific peculiarities[255], and storms which display an especially large number of unusual forms have been reported. Exceptional discharges do occur from time to time, but many reports of highly unusual flashes, as once noted by an authority over 75 years ago, have often been well known for some time to those knowledgeable in the field.[519] Spiral lightning[347] and flashes traveling in very curved and complex paths,[112] sometimes near the ground,[219] have been noted (Fig. 4). The observation of what was evidently a straight, flat segment of lightning traveling horizontally just above the ground, approximately 7.5 (25 ft) long, 40 cm wide, and 5 cm thick, has been reported.[409] A photograph of an apparently similar lightning flash which was also observed, although not in detail, is known.[91]

Chapter 4

The Question of the Existence of Ball Lightning

The most frequent question encountered in the long history of the study of ball lightning is not how the ball is formed or what its properties are, controversial as these problems may be. It is, rather, whether ball lightning really exists. Even following Arago's discussion of this question in 1838, to the present day a skeptical view has been repeatedly expressed. The obstacles to direct experimental study by well established scientific methods and the failure of theories to provide either a satisfying or a conclusive explanation account in large part for the persistent skepticism.

This attitude on the question of ball lightning is not a unique problem in the history of science. The fall of meteors to earth was long considered a superstition of ignorant peasants. Despite repeated observation of these fiery bodies the controversy at one point caused the removal and destruction of rare meteorite specimens from museum collections on grounds they were fraudulent objects of superstition. A thorough analysis of the question in 1794 by Chladni, a physicist whose major work was in acoustics, contained the conclusion that such objects do not originate on the earth and that they indeed fall from the sky. Chladni's study was based on observations by reliable witnesses and data on samples of meteorites, many of which were entirely unlike all ther materials in the area where they were found. His result was, however, not widely accepted. The reality of the phenomenon was finally established by the appearance of thousands of stony meteorites in 1803 at L'Aigle, France. The reports of many reliable witnesses and a great number of actual specimens were cited in the authentication of the event by the physicist Biot for the French Academy of Sciences.

In the negative view of the existence of ball lightning the reported observations are ascribed to mistaken identification of other luminous natural objects and to optical illusions.[224] Meteors are often held responsible for supposed appearances of ball lightning. Several reports originally identified in the literature as ball lightning appear to have been meteors.⁽⁶⁵⁾ The flight of meteors, however, is almost invariably seen as a straight line, in contrast to the characteristic path of ball lightning which is often curved. Ball lightning furthermore appears during storms with very few exceptions, while meteors are observed only by a great coincidence at such times. An ordinary lightning flash seen by an observer directly in its path may appear to be a ball. In the optical illusion which may result, the intense light from the flash persists as an optical image even when the observer changes his field of view. Thus it was suggested that the false image of the ball appears to follow a complex path.

Arago in the first comprehensive discussion of ball lightning took note of this problem.[16] In addition to the presentation of a number of evidently reliable observations he pointed out that an observer viewing the descent of the ball at an angle from the side is not subject to the optical illusion just described. Arago's arguments were evidently effective with Faraday, who in rejecting theories that ball lightning is an electrical discharge took care to state that he did not deny the existence of these globes.[147] The editor of the German edition of Arago's complete works, however, had the temerity to insert a footnote with the categoric remark, "Ball lightning is the result of the action on the retina of the intense light of ordinary lightning" in Arago's chapter on this form of lightning.

Fifty years after the first publication of Arago's review of this problem the persistence of the image of ordinary lightning traveling directly toward the observer was again proposed,[541] and Lord Kelvin at the meeting of the British Association for the Advancement of Science in 1888 commented that ball lightning is an optical illusion from a bright light.[511] The uniform size reported in many cases of ball lightning was ascribed to an illusion associated with the blind spot of the eye.[455]

A direct discussion between opponents in this long controversy took place during a meeting of the French Academy of Sciences in 1890. A large number of luminous globes resembling ball lightning appeared in a tornado which was the subject of a report to the Academy.[148] The glowing spheres entered dwellings through chimneys, bored circular holes in windows, and generally displayed the highly unusual behavior ascribed to ball lightning. Following the presentation of this communication a member of the Academy commented that the extraordinary properties attributed to ball lightning should be considered with reservations since it seemed the observers were suffering from optical illusions.[306] In the heated discussion which followed the observations which had been made by uneducated peasants were declared of no value; whereupon the former Emperor of Brazil, a foreign member of the Academy attending the meeting, remarked that he too had seen ball lightning.[5, 453]

The incorrect identification of St. Elmo's fire as ball lightning has been given as an explanation of many reports of glowing spheres in nature. St. Elmo's fire is a relatively common luminous globe formed by a corona discharge from a pointed object in the ground such as a stake. It appears when the atmospheric electric field is intensified, as it is during a storm. In unusually high fields, which often occur on mountain peaks, this form of discharge may be seen on any protrusion above the ground, even on the hands and heads of people. An explanation of moving globes in terms of St. Elmo's fire, however, requires that the electric field move continuously from one object to another capable of acting as an electrode for the discharge. The progressive motion of a cloud with its associated field along a row of fir trees was suggested to explain a ball which had been reported moving above the row of trees.[197, 268] In apparently similar cases the flight of glowing spheres through the air from an initial point of attachment was also classified as St. Elmo's fire by proponents of this theory.[224] Since the corona discharge requires an electrode, the separation of the globes from direct connection with a grounded point indicates that some other phenomenon is involved, possibly a change in the form of the discharge. There are several reports of balls of fire initially fixed to electrode-like points which later moved freely in this manner.[204, 326, 468]

Other glowing objects are observed in nature and occasionally reported as ball lightning. For example, the will-o'-the-wisp, a nocturnal hunting owl whose feathers are covered with the luminous decaying matter from his nesting place, flies an irregular course above the ground in search of prey. An observer at a distance may report this as ball lightning.[224]

The alternate identities possible in specific cases of ball lightning supply an effective argument against its existence. A leading investigator in high-voltage research has commented that in many years of panoramic photography and observation of storms he has never seen ball lightning.[43] In addition, his direct discussions with supposed witnesses of ball lightning always showed that the observations could be explained in terms of some reasonable alternative. The recurrence of such arguments expasizes the importance of detailed and well recorded observations of ball lightning in nature.

The collection and evaluation of observations has thus long been a major aspect of ball lightning studies. Such collections attempt to fill the role in this difficult problem which duplication of laboratory experiments ordinarily fulfills in science. In some instances observations by reliable witnesses have led to the reversal of a skeptical opinion on the reality of ball lightning originally based on questionable reports and the absence of convincing theoretical explanations.[460, 461, 510] A notable exception is that of Humphreys, a leading American meteorologist. In initial editions of his book on atmospheric physics[223] Humphreys stated that the number and excellence of ball

The Existence of Ball Lightning

lightning observations exclude the view that it is an optical illusion. Consideration of 280 reports which he collected by personal inquiry caused him to reverse his opinion completely,[224] and in the final edition of his work the traditional negative view is forcefully expressed. Humphreys decided that each observation could be conclusively explained by one of the alternatives mentioned previously or by "fixed and moving brush discharges."

Humphreys did not publish the complete collection of his reports. While there may indeed have been many doubtful cases, recent surveys containing hundreds of observations include numerous examples of ball lightning exhibiting the unmistakable characteristics associated with it and not with the reasonable alternatives.[321, 430] Humphreys' requests for information did not make clear to the majority of his respondents with which phenomenon he was concerned. His classification "stationary brush discharge" was evidently meant to include St. Elmo's fire; but "moving brush discharge," if not intended solely as a term for St. Elmo's fire in motion, may be the very form of electrical discharge which has been suggested to account for ball lightning (the theory initially presented by Snow Harris over a century ago which has, in one form or another, found support to the present day). Some observations were rather simply dismissed by Humphreys as optical illusions involving a persistent image, such as that reported by Loeb of typical ball lightning which descended to earth, bounced up, and disappeared as lightning flashed and thunder sounded.[224] Three decades after Humphreys' report Loeb reaverred his observation of ball lightning.[291] In some incidents the possibility that St. Elmo's fire or retention of a bright image by the eye were involved was specifically rejected by witnesses.[461]

The reliability of ball lightning observations has also been questioned on grounds that eyewitnesses close to the occurrence are liable to be greatly agitated, particularly by the stroke of ordinary lightning with which the appearance of ball lightning is often closely associated. In one unusual case, on the other hand, the witness observed the globe with purported calm, without seeing or hearing the flash of ordinary lightning noted by others which struck outside his field of view only fourteen meters away.[546] An unusual absence of thunder or the production of only a muffled sound has been reported occasionally by observers close to the point at which ordinary lightning strikes even while those at moderate distances hear a tremendous crash.

One of the most frequent criticisms of the observations which are an essential basis of our knowledge of ball lightning is that only people completely lacking in scientific training have seen these mysterious globes, or, even further, that no professional observers of the weather[461] and no authoritative investigators of thunderstorm electricity[296] have ever seen ball lightning. This opinion, strongly reminiscent of the debate in the French

Academy over three-quarters of a century ago, is completely incorrect. In addition to the report by Loeb (who would certainly be considered a qualified Electrician according to the meaning of the term in Franklin's time) the appearance of ball lightning was observed from a distance of thirty meters by a scientist[131] from a German laboratory for atmospheric electricity and another by a staff member[504] of the Tokyo Central Meteorological Observatory. Other incidents have been viewed by a meteorological observer,[1] physicists,[124, 177] a chemist,[128] a paleontologist,[305] the director of a meteorological observatory,[441] and several geologists.[499] Astronomers, among those in all scientific fields, have witnessed and reported the largest number.[48, 445]

In rare cases of the occurrence of ball lightning the observer obtained a photograph which recorded the object he was viewing. As in other aspects of ball lightning studies the evidence provided by such photographs is often given inadequate consideration. A discussion of the available photographs is presented in a later section.

The information at hand has led most meteorologists[546] to disagree with the skeptical opinions on the reality of ball lightning which were expressed by the leading authorities Humphreys,[223] Malan,[296] and Schonland.[461] There is no doubt, on the other hand, that many scientists, perhaps a majority of those in other fields, hold the negative view apparently as a result of the unavailability of the data on ball lightning as well as intuitive skepticism. The data from observations are of primary importance to scientific study leading to a more satisfactory understanding of this phenomenon.

Chapter 5

Observations of Ball Lightning

The collection and analysis of eyewitness reports of ball lightning have long been a primary method of study. While the number of observations represented in the literature in individual reports and collections is near one thousand and contemporary requests for additional examples invariably bring numerous responses, this display of nature occurs so irregularly that the scientific methods which have been so successful in providing detailed information on zigzag lightning give no information. Indeed, ordinary photographs of ball lightning are rare. Well established observations, indicated by the presence of more than one reliable witness and good viewing conditions, are considered of great value. Many are republished repeatedly, and because of inadequate identification of the source some appear more than once as separate incidents in collections. An attempt has been made in this work to list recognized instances of republication in a single reference citing each publication as an alternate source. Careful evaluation of each observation is required, for as Brand and Humphreys found, many cases prove spurious as the result of optical illusions or confusion of the object viewed with other meteorological phenomena; and, of course, the veracity of witnesses is involved. The repetition of significant properties from case to case, even with the occurrence of individual behavior which is important, and the existence of photographs showing reported characteristics may be adduced in support for the observational information.

A. Spherical Shapes Associated with Lightning Flashes

The appearance of almost all ball lightning during the activity of thunderstorms and often in close association with ordinary lightning, either directly before or after a lightning flash, presents the possibility, often assumed, that two different forms of the same material, perhaps easily interconverted, are involved. Several instances of unusually persistent lightning flashes which were transformed into a chain of glowing spheres in the lightning channel have been reported. The decay of the flash to bright granules commonly known as bead lightning has been noted since Arago's time.[16, 314, 341, 426] A few cases of bead lightning were observed by investigators of storm electricity themselves, including Planté.[230, 389] A detailed description is available from a lightning flash seen at close range.[40] During a particularly severe storm over Johannesburg a very bright flash struck the ground approximately 100 m from the observer. He estimated that it persisted at least a second and seemed to be 30 cm wide. After the flash died away, there remained in its path a string of 20 to 30 bright luminous beads approximately 8 cm in diameter (one-quarter of the width of the flash) with a distance of 60 cm between beads. The chain was motionless and remained visible for approximately 0.5 sec.

This exact process has been fortuitously recorded in a unique motion picture sequence of a lightning flash which occurred when a plume of water 70 m high rose from the detonation of a naval depth charge.[68] Some frames selected from this film record are shown in Fig. 5. The film shows the initial lightning, evidently a zigzag flash, slowly decaying into separate beads of

Fig. 5. Bead lightning remaining after a linear flash. [M. Brook *et al.*, *J. Geophys. Res.* **66**, 3967 (1961).]

Observations of Ball Lightning 25

light in the original lightning channel which are then gradually extinguished. In contrast to the usual millisecond duration of lightning strokes, the radiation in this observation continued for some tenths of a second, and three distinct strokes occurred in the same path, each slowly decaying to the distinctive separate granules.

The conversion of a continuous lightning column into separate bright regions has been attributed to the pinch effect and to a radial "sausage" instability,[529] processes derived from studies of high-temperature plasma. A photograph[281, 313] presumably of this lightning form is shown in Fig. 6.* The photographers did not see the event which gave this photograph; they exposed the film with an open camera lens for an extended time during a thunderstorm and later inspected the developed film, a common practice in storm photography. The unusually straight line given by the axis of the adjacent luminous segments indicates that an ordinary zigzag lightning flash was not involved.

The conversion of the bead form to typical ball lightning, for example, by the disappearance of all but one bead which then moves in a relatively complex path, has not been reported. It appears that despite the resemblance between the two forms, the occurrence of one excludes the other although the number of bead lightning observations on which this conclusion is based is small. The appearance of a single ball a few seconds after the disappearance of a bead lightning and in the same path was reported in a unique case.[51] The observer at first noted cloud-to-cloud strokes only, occurring at great frequency, 200 or more strokes per minute. A cloud-to-ground flash was then seen with several successive strokes in the same channel. The final stroke appeared especially intense, and it decayed to a chain of beads in the same channel. A few seconds after the beads had disappeared another stroke followed the same path, and shortly after, a glowing ball formed at the upper end of the channel under the cloud layer and floated slowly to earth. Following another strong stroke a bright mass of light appeared, but its descent to earth was too rapid to allow its identification as another ball. The slow motion of a luminous sphere along the channel of a lightning flash immediately after the disappearance of the continuous discharge was observed a few times in another storm with frequent lightning[103].

The formation of a bright sphere on the advancing end of the discharge has been reported in singular cases.[103, 245, 591] Photographs of lightning strokes presumably undergoing this process[38, 361] do not show a curly distinct region at the end of the stroke (Fig. 7) although in one case several witnesses reported the conversion to ball lightning.[38] Two similar reports are available in which the observers saw the complete process of formation of

* Figure 6 is to be found opposite p. 54

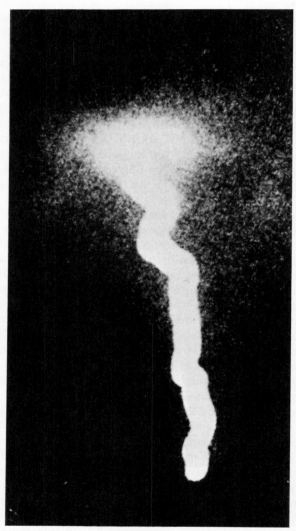

Fig. 7. Round-Tipped Lightning. [H. Norinder, *Kgl. Vetenskapssoc. Arsbok* 94 (1939).]

a ball from a linear flash. In one the witness noted a snake-like discharge descending extremely slowly to a telegraph pole during a heavy rain.[245] It came to a halt approximately one meter above the pole and formed a ball one-third meter in diameter which exploded with such a crash that several people rushed out of doors to see what had happened. No damage to either pole or wires was evident. In the second case the observer selected a specific cloud for observation during a storm.[256] He suddenly saw a white curved lightning descend from the cloud. Its diameter increased as it approached the earth, and it became a large ball of fire. The ball moved slowly toward the earth, and when it was approximately 2 m above the ground and 50 m from the observer it exploded loudly. Observations of bright lightning flashes of this type at short range are considered especially liable to involve optical illusions as discussed previously. However, the appearance of spherical forms associated with lightning activity in storms is supported by these reports.

B. Individual Accounts of Ball Lightning

Study of the ball lightning reports recorded in the literature shows a great diversity in detailed aspects of each case which may be due to the varying and irregular conditions in which the ball occurs. Certain characteristics are noted repeatedly, however, and there is sufficient uniformity to provide a group of typical properties of ball lightning. Since no conclusive explanation of this phenomenon has been given, some investigators hold the view that the unique details in each case should be given in collections lest a clue vital for an eventually successful theory be omitted. In more recent studies of this problem, statistical and tabular summaries have replaced the complete collections of observations which early workers considered essential. At the same time wide variation is evident in the summaries. Many of the complete descriptions of a ball lightning occurrence are unusually clear in presenting the characteristics of this phenomenon, and outstanding ones have reappeared in numerous collections. A series of descriptions selected for the information it provides on basic questions is presented here.

1. A number of observations are available in which the presence of several witnesses or the circumstances under which the ball was viewed support the accuracy of the reports.

Butti, a marine painter of the court of the Austrian empress, sent Arago the following observation of ball lightning.[81] Butti was staying on the second floor of a hotel in Milan in the summer of 1841. A heavy rain was falling accompaned by bright flashes of lightning. The artist was sitting smoking and looking through an open window at the rain, which an occasional sunbeam turned to threads of gold. Suddenly from the street, which had been deserted during the storm, he heard shouts of "Guarda!, Guarda!" He ran to the window to look out toward the commotion and saw a globe of fire at the level of his window in the middle of the street. The shouts came from eight or ten persons who were following the ball at a rapid pace as it moved upwards. It floated quietly past his window, and he turned his head to follow its progress. In order to avoid losing sight of this startling object he hurried into the street to join the crowd following it. Its steady ascent brought it in a few more minutes to the cross of a church steeple, which it struck as it disappeared with a sound like the discharge of 36-lb gun heard at a distance of some twenty kilometers with a favorable wind. In size and color the ball resembled the winter moon in the Alps. It was reddish-yellow, more red in some parts, and it was surrounded by light in which its borders could not be distinguished.

In an incident reported[319] from Wisconsin in 1888 a ball approximately 30 cm in diameter floated into a second-story window in a log house. It went past a young lady in the room, down the hall, and out a window in a room at the other end of the hall as a servant girl ran screaming from it.

In a suburb of Calcutta in 1889 an observer outdoors noted an intense lightning flash near a house which knocked off bricks and caused considerable damage. A group of eight people in a room inside saw a 15- to18-cm-diameter ball of bright yellow fire.[320] About half-way across the room the ball seemed to pause, and then it burst with a deafening report. A stifling orange gas mixed with clouds of dust was formed. The witnesses noted no preliminary signs, no sounds of electrical discharge, or odor of ozone; but the odor of the orange gas was later compared directly to that of a strong nitrogen dioxide–air mixture. The close relationship between an ordinary lightning flash and the fireball observed would lead some investigators to classify this occurrence as an optical illusion.

A very heavy storm in Sweden in 1896 had passed its peak and was beginning to recede, as indicated by the increasing time between lightning flashes and thunder, when a group of six people at the lunch table suddenly saw a luminous white ball floating over the center of the table.[211] They shouted together, "Look at that!" At the same moment the ball exploded loudly with a bright flash, throwing all the witnesses sharply back against their chairs. The hair and clothes of one of those present were burnt slightly, but no one was hurt. All present agreed in this description of the occurrence. A door to the covered veranda and a window were open. The observers felt as if lamed for a time, a symptom noted in many victims of ordinary lightning. Servants elsewhere in the house felt various electrical effects; one fell to the floor but saw no light, and another saw flames. A moment after the explosion of the ball an ordinary lightning flash split two trees outside the house.

Three witnesses saw a fireball[447] during a storm in Karachi in 1895.

The three were sitting in an entirely closed room, and one went to the door to open it. When he turned to sit down again, he saw a fireball the size of the full moon in the air between his friends. Immediately there was a cannon-like crash of thunder. Two of the witnesses felt slight effects, one a sharp pain on his face and the other a trembling arm. There was the strong odor of sulfur. A rifle in its holster in the next room was smashed, and there was a hole where it touched the wall. There were two additional holes in the same wall on the upper floor.

In the summer of 1943 in the United States an engineer finished taking a shower on the second story of a wooden house during a short rainstorm.[542] He heard thunder and saw a 30- to 45-cm-diameter ball, bluish in color, float through a window screen at the end of the hall. It passed through the 9-m-long hall in 3 to 4 sec and then went out a screened window at the other end of the hall. As it floated by him at waist height he felt no heat but smelled the odor of ozone.

2. Some ball lightning appearances have been seen by scientists especially qualified by their field of study for reliable observation of such events.

An astronomer leaving his observatory saw a straight blinding white flash evidently projected from the ground.[48] At the top there was a deep-red glowing ball with a halo which was in motion and extended far from the well-defined borders of the ball itself. The ball emitted a line of fire which was again white and one-tenth the diameter of the initial ray. The new emission returned to the ground, traveling in and out of the initial vertical column twice before finally hitting the earth. The whole sequence was visible for 3 to 3.5 sec.

The director of the Blue Hill Meteorological Observatory of the United States[441] saw a ball of fire fall from the top of the Eiffel Tower at the same time a white flash of lightning struck during a violent storm in 1903. The ball, somewhat less bright than the initial lightning, fell slowly in the interior of the tower from its top to the third platform. The fall of some 100 m took approximately 2 sec. The ball, which seemed to be 1 m in diameter, then disappeared. A visit to the tower the next day established that it had been hit twice by lightning when no one was on the top, and no one there had seen the ball. The tower guard said that such balls had been seen during several storms.

Ball lightning was seen at a distance of 30 m in 1951 by a scientist of an atmospheric-electricity research institute in Germany.[131] The ball appeared outdoors after several lightning flashes during a storm of medium strength. It moved downward at an angle at approximately 50–100 m/sec. It resembled an auto headlight. At the same time another observer in the same room from which the first was looking out saw on the wall opposite the window a four-cornered light as if an auto were passing with a very bright spotlight shining on the window. The ball disappeared behind some trees, and there was directly a loud bang like that from a grenade thrower. A blue cloud of smoke arose which was quickly scattered by the wind above the trees. The ball was seen at the moment of its explosion from another house where the especially bright light had attracted attention. Here bright rays were noted spreading from the explosion flash upward to about 10 m, followed by the blue smoke. The appearance of the ball evidently followed a stroke of lightning to a nearby mast carrying a 5-kV line. Some 5 m from the iron mast a spot 50 cm in diameter was found burned in the meadow grass. Some insulators on the mast were damaged, and a piece of wood was torn off; but no ashes were present in the grass, indicating that the burned spot was not caused by a wood fire.

An observation reported in unique detail was made in 1967 by Dmitriev, a chemist with experience in plasma studies.[128] The ball appeared after an intense flash of ordinary lightning and passed over the observer, who was camped on the banks of the Onega River in Russia. It was seen first moving approximately 1.5 m above the surface of the water, resembling a large electrical discharge. A similar case of ball lightning over a river was witnessed by three members of a geological survey expedition[499] in Canada in 1918. The glowing mass contained a bright yellow-white center 6 to 8 cm in diameter surrounded by two outer shells. Next to the bright core was a dark violet

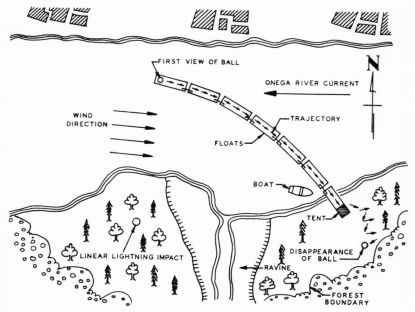

Fig. 8. Path of ball lightning occurrence.

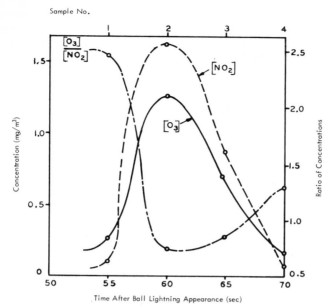

Fig. 9. Atmospheric analysis after passage of ball lightning.

layer 1 to 2 cm thick, and outside this lay a bright blue shell 2 cm thick forming the outer boundary of the ball. The entire mass was somewhat oval in shape, a few centimeters longer than it was wide.

The ball moved steadily at first along a line of floats in the river which protruded 10 to 15 cm above the water. This path took it somewhat at an angle to the prevailing wind toward the observer on the bank (see Fig. 8) It moved at 1.5 m/sec, or slightly faster, and rose steadily a few centimeters for each meter passed. The fireball flew over the observer's head and then toward the forest lining the river bank behind him. Over land its speed decreased to an estimated 0.4 m/sec while it climbed more rapidly at 0.5 m, or more, per meter traveled. It remained motionless for 30 sec over a knob of land.

The witness reported taking four samples of air soon after the ball passed over him with evacuated gas sample bulbs which happened to be ready for use. The samples were collected with the bulbs held at arm's length as high as the observer could reach toward the path taken by the ball. Analysis of the gas collected indicated the presence of both ozone and nitrogen dioxide in amounts much greater than normal in air, as shown in Fig. 9. Other gases usually found in air were unchanged in concentration. The observer noted a strong odor at the time resembling that produced by high-energy radiation in air. A transistor radio in the observer's camp gave increasing static in the early stages of the incident, which soon became a continuous rumble of increasing loudness. The ball itself was heard crackling loudly at close distance by the eyewitness. It left a trail of bluish, acrid smoke which slowly dissipated. As it approached the thicker forest, the globe collided with several trees and emitted violent sparks in six or seven such encounters. The collisions caused it to zigzag, and its color changed from the original white to bright red. After remaining in view for 60 to 65 sec, the ball disappeared.

From his observations and measurements during this incident the witness attempted to derive additional properties of the fireball. He suggested that the ball was formed by a short section from the preliminary flash of ordinary lightning which separated when an instability in the lightning channel, a sausage plasma instability, caused complete contraction of the lightning in two places and separation as a globe of the short length thus cut off. If the material of the fireball originated in the preliminary lightning flash, its total lifetime was approximately 80 sec, of which it was observed directly by the eyewitness for 60 to 65 sec. From his experience with plasma characteristics the viewer compared the strong intensity of the yellow-white central ball to a plasmatron torch with a temperature of 13,000 to 16,000°C. He estimated the radiation which would produce in air the odor of ozone and nitrogen oxides left by the ball at 1,000 to 30,000 roentgens. Although the fireball evidently passed very close to the witness, he did not report any direct sensation of the heat which might be expected from a globe of such high temperature; and a film detector which would indicate radiation showed no effect although it had been 2 m from the ball for approximately 6 sec. A meter for γ-radiation indicated a radiation level of 1.2 milliroentgens per hour, but the witness pointed out that the radio-frequency radiation detected on the radio could have affected the photomultiplier to give an incorrect reading. The formation of ozone and nitrogen dioxide by experimental electrical discharges in air was compared with the amounts of these gases found near the path of the ball lightning. The ratio of ozone to nitrogen dioxide near the experimental discharges decreased with increasing discharge voltage. The highest ratio measured in the trail of the lightning ball as shown in Fig. 9 was 2.5, which corresponds to a potential of 300 to 400 kV to ground if the lightning may be compared to a silent discharge. The ratio of hydrogen to oxygen found in the samples collected in nature was far below that needed to give an explosive mixture by a factor of 10^{-3}. The stability and relatively long life of the glowing ball were ascribed to a cool outer surface of negative molecular oxygen ions formed by attachment of electrons initially present to atmospheric oxygen. This layer was presumed to retard diffusion of charged particles and heat transfer from the ball into the surrounding atmosphere. These aspects of interpretation of the observations made during the incident and presentation of a ball lightning structure are discussed in later sections dealing with plasma characteristics of ball lightning and theoretical models.

3. Many ball lightning appearances are closely related to storm phenomena; but there is no uniformity, for example, in their formation, necessarily resulting from the relationship. Some balls have been seen dropping right out of clouds, others are formed directly after a flash of ordinary lightning, and yet others are immediately followed by lightning. Some evidently are not connected with other events in a storm at all.

Several balls of fire dropped into the sea from a large black cloud over Sussex in 1780. The observer was than struck by zigzag lightning which paralyzed him temporarily while others in the house were struck dead.[67]

Great lightning activity during a storm in France in 1903 attracted an observer to his window.[407] Soon after a very intense flash and loud thunder, he saw a firework-like meteor flying from the line of the houses on his street where the previous flash seemed to strike. It was a small yellow mass leaving a luminous trail behind it, first in a horizontal path and then gradually descending, as if from gravity, in its flight. It disappeared in approximately three-tenths of a second without exploding. The witness was certain it was a lightning ball.

A large ball of fire was generated when a direct flash of lightning struck an antenna during a heavy rainstorm in France in 1910. The antenna had been erected to study electricity in storms and two previous lightning discharges in the storm were shown on the recorder. A third flash struck the antenna and the observation post, vaporizing 65 m of the 2-mm-diameter copper antenna wire. The resulting ball of fire was seen by several witnesses in a sawmill nearby who were awakened by the thunder of the storm.[528]

A bright yellowish ball evidently emitted from a sharp bend in a very bright, branched lightning flash was seen by an observer[175] with watch in hand in 1927. The flight of the ball started approximately 1 sec after the lightning stroke. It followed a straight path for one more second, and in a short time some thunder was heard followed by a single, loud detonation. From the time between the disappearance of the ball to the explosion the observer estimated the distance of the explosion at 1,150 m. A few hours later it was learned that the ball fell on a small house 1,100 m away. Many witnesses described its fall similarly. The ball was not seen inside the house, but nearby the tip of an electric power pole was shattered. Electric fuses were burned out in many houses, and the wall plaster fell in the house struck by the ball. An estimate of 1,200 m/sec was given for the velocity of the ball lightning from the time measurements.

A series of almost vertical and straight lightning flashes was noted by two observers during a violent storm in France in 1901.[544] After a lapse of some minutes, both saw a ball of fire falling like a stone from the skies in the same place where the linear lightning had been and from the same height. After another interval the same region was illuminated repeatedly by diffuse localized discharges.

Four men had resumed their work in construction of a wall following a rainstorm in Germany in 1868. The sun was shining, and in the blue skies there were only a few almost transparent clouds. The men were about to lift a stone approximately 80 cm square to the wall. They were standing around the stone when suddenly there was a flash of lightning; in the middle, approximately 90 cm above the stone, a round, yellow, transparent ball of about 20 cm diameter appeared, steadily moving up and down for a distance of 4 cm. In the center of the ball was a bluish flame which was pear-shaped with the point downward and 4 cm in length. The flame revolved in a vertical circle of 7 cm diameter inside the large ball. A sharp crack was heard after a few seconds, and the ball lightning disappeared. Another workman in an enclosed area 10 m away didn't see the ball but heard the explosion and rushed out. Soon after, they heard that it had struck in a quarry 100 m away.[140]

4. Many reports indicate that the appearance of ball lightning is accompanied by the presence of strong electric fields.

A traveler noted four separate storm centers, three of which displayed a great variety of lightning discharges with great frequency.[163] The fourth gave only a few glowing white strokes to ground with long intervals. Suddenly from the middle of a cloud which had been marked by its lack of activity while lightning was especially strong all around it a blinding white ball approximately 20 cm in diameter emerged and went straight down, slowly and steadily. Halfway to the ground it burst, and six glowing parts now moved in divergent directions, disappearing almost at the same time in the last third of the way to the ground. The fragments seemed to travel slower and glow more faintly than the initial body. The entire process lasted 10 sec. No explosion was noted by the witness, possibly because at the moment of fragmentation the telegraph wires near him hissed and crackled.

In another incident a hissing sound from an upper window, which was open, attracted the attention of a man working at a desk in 1927 in Switzerland.[325] When he looked up, he saw a red, glowing flame 1.5 m above him. It went by him like a rocket and burst with a loud bang in the middle of the room 2.5 m above the floor. Another witness 4 m away saw it through an open door. The light fuse was burnt out, but except for a sulfur smell there were no other traces. Outdoors it was cloudy, and rain started approximately one-quarter of an hour later. A recorder approximately 600 m away in an observatory indicated strong atmospheric disturbances at the time. The potential varied greatly to values as low as -290 V/m. While this was normal for a storm, the conductivity did not change correspondingly, nor were the expected storm effects in temperature, pressure, or humidity observed. No specific storm discharges were seen at all.

An exploding ball lightning occurred[438] at the end of a short violent storm in France in 1904. The ball which was seen at three different places separated by 500 m, all in a straight line, burst with a loud noise. Three large chimneys were damaged in its path, and strong electrical effects were noted in general. Several people felt electrical shocks, door bells began to ring, and flame-like discharges were seen inside houses. A great disturbance on the wall of a city building only 30 m from the starting point of the ball occurred despite a lightning rod which was later found to be in good condition. It appeared that the rod had no effect on the lightning ball.

The formation of a small fireball in an unusual incident during a thunderstorm was seen at close range.[224] A lady was standing on a rug during the storm with her hand at her waist and one finger slightly extended. A witness 1.5 m away noticed the air between her finger and the floor wavering as if warm. Something rose slowly from the floor up to her finger, and a pecan-sized oblong fireball appeared attached to her finger for a moment. It seemed to glow through a haze. The ball disappeared as lightning flashed outside. A similar case occurred on a fly swatter during a light storm.[336]

5. Several types of motion have been observed in ball lightning; for example, the glowing masses often roll or bounce, rotation is frequently reported, and others travel in a consistently upward direction.

During calm drizzling weather a 15- to 20-cm-diameter lightning ball was seen in 1956, after it rolled down a tree.[384] No ordinary lightning was noted during this time. The tree was marked by a very weak trace unlike that often produced by zigzag lightning. The ball appeared bluish-white at 300 m from the observer, then red when at 2 m. It was seen falling slowly at an angle of 30°; it then bounced off the wet meadow up to 4 m and moved slowly along a barn. It circled a tree and hit a fence post with a loud explosion. Although a splinter 10 cm long and the thickness of a finger split off the post, leaves near the post were not disturbed, and there were no traces of burns near the spot. After the ball disappeared, zigzag lightning hit the barn igniting a fire.

During a short violent storm in Germany in 1905 a bright ball lightning approximately 20 cm in diameter formed.[265] After a moment in which it remained stationary, it moved in a straight line downward, accompanied by a faint noise like that of an electrical discharge. On its left side were small portrusions from which shadows moved over the surface, indicating a rotational motion. After 5 or 6 sec it exploded illuminating the nearby region with a red light.

A hail storm, accompanied by thunder, on the French coast of the

Mediterranean in 1877 was followed by a few drops of rain mixed with partially melted hail; but later the sun was very hot, and there was a clear sky at sunset.[49] Then, in the clear sky, numerous lightning strokes were seen with no sound of thunder. In the east a layer of black clouds appeared with one particular mass in active motion. Balls of fire shot out in all directions like fireworks, apparently from a point in its center. The red or sometimes yellow balls became white and burst silently with a blinding flash after traveling 6–10°. The diameter of the balls at 18 km distance was 1°. They seemed like large, very light soap bubbles moving slowly in horizontal paths parallel to the plane of the clouds. None were directed downward. They appeared three or four times in 2 min. A hail storm with frequent ordinary lightning followed. A similar display was reported by residents of the area several years earlier.

A yellow ball of fire approximately the size of a large cricket ball rose from behind some houses near London Fields, 1.5 km from the witness, in 1874 during a fearful storm with wind, rain, hail, and lightning.[526] The ball rose slowly at first but accelerated as it ascended. The increase in velocity was such that when it turned sharply at an elevation of 45° it produced the appearance of forked lightning. It zigzagged three times at it rose into a dark cloud from which flashes of lightning were seen. Approximately 10 min later another ball rose in the same way although not from exactly the same position.

 6. Very rarely two balls linked together have been reported.

Two orange globes linked vertically by a slightly luminous granular string were seen[125] by four witnesses in Germany in 1912. All agreed that the part of the sky in which the balls appeared was free of lightning during the whole storm. The two masses appeared suddenly approximately 100 m above the ground and 1 km away from the viewers. The two balls were separated by approximately 1.5 m and traveled northeast at 1 km/sec keeping their centers on the same vertical axis. The wind at the observing position came from the north. As the larger upper ball traveled horizontally, the lower one fell slowly, increasing the distance between them and apparently stretching the connecting filament, which disappeared. The lower ball fell behind a thicket. The distance and its dimensions as observed indicated that its diameter was at most 1 m. The lower globe was visible for 0.75 min; the upper one disappeared suddenly after 2 min as it began to fall slightly.

Two balls of fire linked together were also observed[196] during the eruption of Vesuvius in 1794. They flew close to the house from which they were being observed and seemed to fall, one into the sea with a splash audible to the witness and the other on the land.

 7. Ball lightning has appeared in entirely closed rooms, one of the most interesting problems for theory. Reports originate equally from open spaces, mountain peaks, and at sea. A collection[323] of eighteen ball lightning incidents at sea was published in 1890.

A glowing ball the size and color of an orange was seen at the metal door handle in a closed room at the same time as thunder sounded.[303] The sphere disappeared with a large explosion without moving. There were no traces of burns, but a strong ozone-like smell was produced.

A scientific observatory assistant returning home during a heavy storm with strong lightning was just at his door at the end of an entry hall 20 m long and 3 m wide when a very loud detonation burst right behind him directly followed by an intense light.[377] He turned around and saw some five paces down the hall a very bright, blue–white glowing ball floating down from the hall ceiling. Its outline was irregular and seemed to change, but the intense light prevented detailed observation. The mass descended almost vertically at moderate speed, not more than 2 m/sec. When it reached the floor it disappeared without a trace, noiselessly, and leaving no odor. Close examination showed that the white paint on the ceiling had been blackened, but no sign of a burnt spot could be found on the floor. Several seconds later there was a very intense lightning discharge nearby from which the electric wires emitted sparks, switches were burnt through, and the lamps melted. The witness stated that the later discharge was independent of the fireball. He was uncertain of how the object had come into the hallway, having seen it only on its descent from the ceiling to the floor, but gave the opinion that it had probably entered through a crack in one of the numerous warped windows in the side of the entrance.

A dull red, pear-shaped fireball floated slowly from an enclosed room with 2-ft-thick walls used as a drying oven.[297] Short streamers of red flame flared out from it in all directions. It passed across a wooden landing, over the top of a truck, and exploded 1 m from the ground. At this time rain had started to fall. An observer of this incident, reported in 1937, found that the circuit breaker on the electric light wires tripped and a fire started where the wires were within a few inches of the lightning ground coming from the mill stack on the outside of one of the thick walls of the drying room. There was no sign that anything had penetrated the wall at that point into the room where the fireball started.

A couple climbing in Switzerland in 1885 at an altitude of 2504 m saw a display of numerous globes.[490] Heavy rain and snow were falling, and there was almost continuous lightning. On a ridge they saw suddenly a horizontal row of small yellow globes which then coalesced to form a large luminous mass which itself emitted red and blue globes. These exploded as they fell. Soon after, a single fireball appeared, moving in parabolic curves back and forth with a speed like that of a thrown ball. After seeming to disappear, it started again, continuing for some minutes.

Unusual cases of ball lightning at sea are found in several reports, possibly in accord with the ample opportunity and unrestricted view of weather phenomena. The works of Robert Boyle[63] describe an incident on a ship in 1681 often ascribed by later authors to ball lightning. Lightning struck the *Albemarle* in a storm 100 leagues from Cape Cod. The main top sail was burnt, and the main-cap and mast split. Following an especially loud burst of thunder a burning mass fell from the clouds on the ship's boat, breaking it into several pieces. The mass burned until completely consumed, with a smell similar to gun powder; the men could not extinguish the blaze with water or remove the object with sticks. The solid mass observed may indicate the coincidental fall of a meteor or the ignition of a fragment of the mast, although its descent from the clouds was specifically noted.

The *Aeolus* was sailing to Valparaiso in a heavy snow storm in 1881 when a glowing ball approximately 0.5 m in diameter fell directly from the zenith into the water 2.5 m from the ship.[427] A fearful explosion sounded directly followed by muffled thunder. The two men at the helm were blinded for several minutes. Two sailors on the foredeck reported they were thrown back and struck by the lightning. Two minutes after the explosion, St. Elmo's fire was seen on all three topmasts.

A pear-shaped lightning ball was seen by Marsh,[305] the American paleontologist, in 1878. Marsh was on the deck of a large yacht during a thunderstorm in which there were several lightning flashes in the harbor of Southampton, England. On looking toward a bright light near the upper part of the foremast, he saw a rose-pink ball of fire distinctly at halfmast height falling slowly toward the deck. It was approximately 10 cm in diameter and 15–20 cm long, with the larger end below. It struck the deck and exploded, knocking a man down. A part of the stroke evidently went down a ventilator into the galley, where it knocked a pen from the cook's hands and moved utensils about. There was a strong ozone-like odor which remained for some time. Marsh saw the forward deck shining with bright confused light, and an officer there saw lightning streaks in snake-like motion on the deck.

8. Incidents of the fireballs in airplanes form an increasingly large group which has become significant recently. Numerous discharges of ordinary lightning strike aircraft, and some cases of the glowing spheres are evidently associated with such strokes.[358] Ball lightning appearances in aircraft give evidence of high temperature and explosive action in a number of cases. The reports available in the literature are summarized here.

In 1938 the captain of a British flying boat (BOAC) in flight at 2,500 m in a dense nimbostratus cloud saw a fireball come in through his open cockpit window.[183] His eyebrows and some hair were burned off, and there were holes in his safety belt and dispatch case. The ball traveled through the plane to the rear cabin where it burst with a loud explosion.

A commercial airliner (TWA) flying from Paris to Cairo in 1948 at an altitude of 3400 m was passing through a very cloudy region when a bump was felt under the cabin.[34] A passenger looking out the porthole saw a ball of fire, orange-yellow in color and slightly larger than a tennis ball, rise from under the cabin. A dark gray-violet layer surrounded the bright center. The ball was 2 or 3 cm thick with a short tail, as if it were a rotating spiral, and it was traveling the same speed as the airplane. The observer saw the ball burst approximately 30 cm from the side of the cabin as it ejected a bright ray forward almost 3 m long. This resembled a magnesium light according to a witness in the next seat. At the same time it gave a loud detonation, louder than a revolver. The appearance seemed to last 1 or perhaps 2 sec.

The explosion of a large fireball in a cloudy sky, comparable in appearance to other reported ball lightning, was credited with the complete destruction of a military fighter in flight in 1948. There was no direct observation of the event although several saw the flaming red ball earlier.[42]

In 1956 a Russian LI-2 transport, a propeller-driven plane, was struck by ball lightning while flying at 3300 m in cumulonimbus clouds.[257] The flight had passed through a stormy cold front and bad icing occurred. The cloud peaks were up to 5.5–6.0 km, air temperature at flight altitude was 2–4° below freezing, and the plane was in very rough air. An orange-red fireball 25–30 cm in diameter approached the craft rapidly from the front. When it was 30–40 cm from the nose, it swerved to the left and passed by the cockpit. As it did so, it hit the port propeller in the upper part of its arc and exploded. There was a blinding white flash, and a loud explosion was heard over the engine noise. A flaming stream passed along the port side of the fuselage, and the aircraft rose sharply. Strong radio interference had been noted in the flight through the storm cloud. When the radio operator tried to disconnect the antenna after the lightning discharge, he received an electric shock. The only damage to the aircraft was in a small region of the propeller blade, the trailing edge having been melted over an area 40 mm in length and 5–10 mm in width at a distance of 30 mm from the tip of the blade (see Fig. 10). This region was surrounded by a loose coating of soot, easily brushed off by hand.

A Russian jet plane also encountered a lightning ball in 1956 while climbing at 2500 m in a storm cloud.[257] The flight passed through cumulonimbus with peaks to 4500–5000 m in rain and storm over European Russia. The ball appeared in front and starboard of the plane and burst almost immediately giving a bright flash and an audible explosion as it disintegrated into smaller fragments. The engine near the ball stopped, but the pilot succeeded in restarting it. Inspection after the flight showed no visible damage, and the flameout was attributed to the rarefaction from the explosion.

Another Russian incident occurred in 1959 when a ball of the traditional

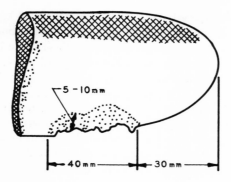

Fig. 10. Propeller blade damaged by ball lightning.

"fist size" (possibly 10 cm in diameter) struck an aircraft over Irkutsk.[397] The plane was flying at 400 km/h in dense clouds occurring in two layers, 400–4000 m and 8000–10,000 m. At 1400 m in a storm the wingtips were invisible, and the air was very rough. Rain and snow quickly coated the front windows and wings with snow. Ordinary lightning discharged toward the ground in front of the plane. At the same time a strong blow jarred the plane, but it remained under control. At this time the fireball passed on the port side of the plane, and a shower of sparks appeared. The pilot took the plane up to 4000 m out of the clouds. The cabin was strongly magnetized, causing a 100° error in the compass. The radio was operating, but the radiocompass was incorrect. Inspection after the flight showed that rivets had melted off on the port side in the front of the fuselage, but the skin itself was not destroyed (see Fig. 11).

An incident occurred in 1960 in a KC-97 Air Force tanker loaded with JP-4 fuel.[533] The plane, one of the type used to refuel short-range aircraft in midair, was flying on instruments in clouds at 5400 m with light precipitation but temperature above freezing. St. Elmo's fire was reported around the front window edges, a fairly common sight. The pilot observed a yellow-white ball approximately 45 cm in diameter enter through the windshield and pass between him and the copilot at the speed "of a man running." He braced himself for an expected explosion. The ball went down the cabin passageway past the navigator and engineer. In approximately 3 sec the fuel boom operator in the rear reported on the intercom that the fireball rolled through the rear cargo compartment and then passed out over the right wing into the clouds. No sound was emitted by the object. The pilot had experienced ordinary lightning strokes in previous flights and expected an explosion from the fuel load. This expectation was also reasonable in view of the other ball lightning events on aircraft described here.

A passenger observed a similar traversal[232] of an airplane cabin by

Observations of Ball Lightning

a glowing ball following an encounter with an ordinary lightning discharge. A flight from New York to Washington (Eastern Air Lines) in 1963 was passing through an electrical storm when it was surrounded suddenly by a loud bright flash. Seconds later a blue-white sphere estimated at 22 cm in diameter came from the pilot's cabin down the aisle of the airplane, passing approximately 50 cm from the observer. It remained at the same height, perhaps 75 cm above the floor, and moved in a straight course as far as he could see it. From this very close view the witness noted several properties. The ball appeared to be moving approximately 1.5 m/sec with respect to the aircraft. It was spherically symmetrical, almost like a solid ball with slightly darker edges and no evident toroidal structure. Its luminosity was equivalent to 5–10 W, and no heat was radiated. No rotation was perceptible, and the ball seemed to be very stable.

These innocuous incidents are evidently relatively uncommon in the airplane cases of ball lightning, although such harmless events may even be in the majority of the usual appearances near the ground. Possibly many harmless cases in aircraft remain unreported.

In another case of ball lightning in an airplane a Russian LI-2 transport was hit in 1965 by a larger ball 60–80 cm in diameter while flying over the Kola Peninsula in thick nimbostratus clouds.[578] Visibility was limited to less than the wing tips. The flight was passing the northern portion of a cyclone. The temperature at flight altitude was $-5°C$. A noise like a rifle shot

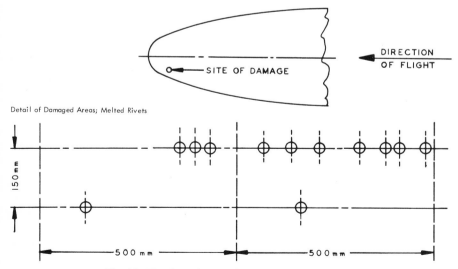

Fig. 11. Fuselage rivets melted by ball lightning.

was heard when the sphere struck the front of the aircraft. The passengers and crew saw red streaming discharges moving along the surface of the plane. Similar effects were noted in the 1956 case involving the same type of aircraft [257] and a shipboard case in 1878 in which snake-like streamers were reported on deck following the explosion of a smaller ball.[305] Following the collision of the airplane with the ball the radio compass rotated, and the magnetic compass spun erratically for 3–5 min. When the latter stopped moving, its readings were incorrect. All radio apparatus was also disabled. Inspection of the aircraft after landing showed several rivets damaged in the forward fuselage and two holes 1.5–2 cm in diameter melted in the trailing edge of the elevator.

The appearance of ball lightning in aircraft, while rather seldom, is evidently clearly established by these several reports which also indicate potential hazards, e.g., explosions, temperatures high enough to melt metals, strong electric effects, and force sufficient to deflect the craft from its path. The generation of the fireballs in completely closed spaces is further shown to include structures almost entirely surrounded by metal.

9. Ball lightning has been reported in connection with volcanic eruptions and cyclones.

An electrical storm with many strong discharges and strong field effects occurred on the lee side of a volcano in Guatamala in 1902. Corona discharges were seen on people and houses everywhere, and there were many lightning balls which burst with muffled explosions but caused no damage. There was much thunder heard from the volcanic clouds, often with no evident association with lightning flashes. Broad flashes and a few ball lightnings were seen, but there were no normal zigzag flashes observed.[286]

Great electrical activity with continuous intense lightning flashes occurred during the Martinique cyclone of 1891. Ball lightning appeared in all directions during the storm. The crackling globes traveled through the air for several minutes and burst approximately 0.5 m from the ground.[513]

A few witnesses have been able to report on the appearance of fireballs inside of tornado funnels.[550] The absence of ordinary streak lightning was noted in one example in which the discharges took the form of a fiery stream which flowed from the sides of the boiling, dust-laden cloud and broke into spheres of irregular shape as it descended. Fireballs were reported[126] in connection with three recent tornados in France from 1961–1965.

10. According to several reports, the glowing mass of ball lightning can itself be the source of sparks or rays or, in a few cases, a discharge resembling an ordinary flash of lightning.

Frequent heat lightning was seen on a hot day in Paris in 1849 under a calm sky.[142] A large red ball, at first mistaken for a balloon ascension, was seen from a second floor window. While the observer was puzzling over the identity of the object, a fire started at the bottom, which was 5 or 6 m above a tree. At first there were small sparks and flames, evidently through an opening in the ball. When the opening seemed to be two or three times the size of a hand, a sudden frightful detonation fragmented the whole mass. Zigzag lightning flashes were emitted on all sides. One hit a house and made a hole in its wall like that of a cannon ball. Three men in the street were knocked down by the detonation. Some of the initial material evidently remained, as a white flame resembling the light of fireworks continued to burn brightly. The incident lasted for one minute.

A rod-shaped object 13 cm across and 38 cm long, glowing yellow like hot iron, descended from the sky approximately 100 m from an observer.[199] The sky was clearing after passage of a violent rain storm in 1889 in England. The object descended slowly enough so that its progress could be followed, and when it was 12 m from the ground, near a chimney, it emitted a horizontal flash as if it were bursting. The ray was white in the center and red near the outside. There was a violent explosion, and, soon after, a strong smell like burning sulfur was noted by observers. There was no effect on the chimney evident on the outside, but the kitchen filled with smoke and soot as did the dining room. The master of the house, who was entering this room, heard a detonation and at the same time saw a bright flash. Some bricks were thrown 2 m and plaster 5 m in the room.

During a storm in 1949, described as of hurricane strength and lasting only 30 min, in Germany a couple was wakened by a bright light in their room.[171] The husband leaped out of bed and saw through the window, 10 to 12 m from the house and 3 or 4 m up, a dazzling white ball floating in the air. It seemed to be the size of a full moon at its height, and it was emitting rays. After a few seconds the ball broke up, but a part remained, like a glowing new moon curving downward. There was a strong discharge to the ground and an active spark in the air. The rest of the ball, which was still floating at the original height, emitted numerous foot-long reddish-yellow sparks to opposite sides, which were extinguished before they reached the ground. This process lasted at least 4 sec. His wife noted a glowing mass the size of a hand which hit the earth and sprayed out red sparks.

11. In a few cases glowing spherical discharges attached to conducting points which might be classified as St. Elmo's fire are transformed into moving objects with behavior typical of ball lightning. Fireballs are also seen traveling along power lines during storms.

During a storm in France in 1895, in which there was no lightning or thunder, a large fireball was seen on an iron tower supporting a telegraph wire on the top of a house.[326] The ball was approximately 30 cm in diameter, and it appeared bright and distinctly defined to an observer 100 m away. Large sparks were spraying out from it. After 40–50 sec the ball suddenly split into three small globes the size of a child's balloon. The sparks were no longer emitted, and the three balls rolled along the roof as if under the effect of gravity. As they came to the rain gutter all three disappeared without a detonation. Another had reported the appearance of single fireball in the same place. Later it was noted that the wire support was no longer vertical but was visibly tilted.

A strident noise was heard from the lightning rods of the cable office in Martinique in 1899 as a caller was awaiting a telephone connection.[270] This was often observed during the storm season, and the caller had started to put the phone down on a ground wire when he saw a fireball 20 cm in diameter moving along the telephone wire. It was glowing as brightly as a 20-candle-power electric lamp. As it arrived at the receiver, a large discharge occurred between the globe and the ground wire, which was fortunately near the receiver. There was a detonation like a small cannon accompanied by a blinding light. The observer was momentarily stunned and could only ground his line when he recovered in a short time. He found that the telephone was completely burnt out. The relay of the Morse telegraph was damaged slightly; a long length of the coil was melted. The plates of the lightning rod were melted and even fused together in several places.

During a violent storm in 1906, the chief of a railway station in Switzerland saw, with five or six other witnesses, green sparkling globes the size of eggs going along the electric wires toward the mountains.[180] These objects disappeared in the distance, and in a short time they came again, in the same direction and moving very slowly. This occurred several times in one-quarter of an hour. In the same storm, a ball of fire was seen on the electric wires inside the house after an intense lightning flash, and one appeared from the telephone wires in a convent. The last disappeared almost immediately, breaking a window and a heater flue.

On an oppressive day in Scotland in 1947 in which, however, there was no rain or thunder, a fireball was seen[89] running along an outside electric wire. It struck a very large oak with a terrific explosion, shattering the tree to pieces. In the house nearby the radio, telephone, and all fuses were burnt out; but the detonation did not break any windows or cause other damage.

12. Despite the numerous examples in which the luminous balls observed in storms exploded and caused appreciable damage, the phenomenon has often been classified as harmless because of the frequent cases in which it touches people without harming them and disappears noiselessly in a characteristic appearance which can be designated "benign" ball lightning. In some descriptions[465, 518] it has been classified as entirely harmless, potential danger being attached only to the ordinary linear flash of lightning often reported as preliminary to appearance of the ball or occurring immediately as it vanishes.

In 1904, a German engineer who had briefly seen ball lightning at a great distance ten years earlier was walking with his wife in a storm with rain, hail, and snow while a strong wind blew.[50] They saw a large bright ball 4 m in diameter and 6 m above the ice at a distance of approximately 30 m from the road which they were following. The ball sank through the telegraph wires, which glowed, and then enveloped the couple. They stood in a thick white sea of light in which the sensations of odor or heat were absent. There was no breeze from the motion of the ball, and they could not feel the outside wind. They could see only the pebbles of the road. The man's cigar, which was lit, was unaffected. The ball then crossed the road leaving them. It seemed to rise and, at a distance of 10 m on the other side of the road, disappeared in the hail storm. The light radiated by the ball was approximately 34 candlepower, and it seemed to travel by a sliding motion. During the 4 sec in which it was in view it traveled 50 m. An inquiry showed that the telegraph equipment in the post office was not being used at the time, and no disturbance was noted; but a watchman in front of the office said he saw a ball pass over at approximately the same time.

Eleven people were gathered in the salon of the Baron of France in 1900 during a violent storm when a blue ball the size of a child's head appeared in their midst.[79] It slowly crossed the room grazing four people without hurting them. As it departed through an open door in front of the entrance to the grand stairway, it exploded.

13. In other reports dangerous aspects of ball lightning are clearly indicated, including explosions which result in death or serious damage and evidence the release of great energy.

A child who touched a lightning ball with his foot was killed in Germany in 1865. His toys were tossed around by the explosion.[193]

Zigzag lightning struck a tree in Austria[229] in 1907 during a heavy storm. The flash spiraled down to the ground. Two men sitting in front of the nearby house were thrown 3 m. They felt great heat but were not hurt. The storm passed and gave way to a clear sky. Two hours after the flash of lightning a fire was being made in the kitchen fireplace when a severe explosion occurred throwing the masonry about. A fireball the size of an apple appeared before the eyes of the terrified housekeeper and floated into the chimney with a crackling sound.

A traveler was overtaken by a violent storm and took shelter near a stable.[154] He saw two children 12 or 13 years of age who for the same reason were standing under the doorway of the stable in which there were 25 cattle. In front of them was a slope which ran for some 20 m to a large pond where there was a poplar. Suddenly a ball of fire the size of an apple came down from the top of the poplar branch by branch and then followed the trunk. It rolled on the ground very slowly, as if picking its path among the pools of water, until it came up to the doorway where the children were. One of them had the courage to touch it with his foot. Immediately a shattering detonation shook the walls of the farm. The two were thrown to the ground without being hurt, but 11 of the animals in the stable were killed.

Knowledge of the previous incident and others similar to it in the literature may indicate an advantage in untutored caution over academic curiosity in the following ball lightning occurrence. After a severe lightning stroke in a storm, a yellow flame 12 cm in diameter and spinning like a top appeared in a house.[583] There was a strong smell from the ball, not a sulfurous odor but perhaps that of nitrogen oxides, and the ball did not explode. A cook witnessed this appearance from a distance of 1 m before she ran away. Professor R. W. Wood of Johns Hopkins commented, "The cook was near enough to the ball to touch it, and it is regrettable that she neglected the opportunity of making a valuable contribution to our knowledge of this mysterious electrical phenomenon!" He added thoughtfully, "I think that I should have reached for it, but am not sure."

One of the most discussed appearances of ball lightning, on which estimates of its energy content have been based, involved a ball which fell into a tub of water.[184, 345] A small red-hot ball was seen coming down from the sky. It struck the house from which the observer was looking, cut the telephone wire, burnt the window frame, and then fell into a tub of water under the window. The water boiled for some minutes. When it had cooled, the witness inspected the tub, but could find no residue of the ball.

14. The intervention of optical illusions has been suggested to account for eyewitness accounts of ball lightning. Despite the frequency with which this suggestion has been made, reports corresponding to the illusions which have been described are rare and difficult to find.

One incident involved the appearance of a ball which the witness himself concluded was an optical illusion.[287] The luminous globe appeared at the same time as a flash of lightning with simultaneous thunder according to the observer, who was walking in England in 1877. The glowing ball, approximately as large as a parasol, fell in front of him and then rolled along the road in the direction he was walking. It disappeared, evidently by turning in a private gateway. The witness reported he had decided almost immediately that the image was entirely subjective from the effect of the initial flash of lightning on his retina, following the direction of his eye as he walked and disappearing when the effect on the retina wore off.

The illusion most often proposed is attributed to the viewer observing an ordinary flash of lightning while directly in its path. Since he sees only the crosssection of the flash, it appears to be a round lightning mass traveling toward him, assuming that no change in direction occurs, as it does in the common zigzag path. Arago noted that this illusion would not account for cases of ball lightning observed descending laterally across the observer's field of view.

An evidently unique report seems to have dealt with just such a line-of-sight lightning channel although the witness ascribed the phenomenon to ball lightning.[114] The observer was sitting at dinner in a Norwegian hotel in the summer of 1900 watching a vivid display of lightning during a thunderstorm. Suddenly a yellow streak apparently 3 cm broad darted right toward him from the sky as if to strike his forehead. The witness was too spellbound to move, and there was insufficient time even to call out to other diners. Just as the object came to the window it changed to a dazzling yellow ball of fire the size of a cricket ball. It burst with a frightful crash shooting out large violet flames in all directions. The next day the observer found the track of this flash in the ground beginning 20 m from the window through which he had seen the occurrence. Small sods and tree branches covered the surrounding area, in which two or three small trees had been uprooted. The track was a furrow in the ground 44 m long, varying in width from 7 to 24 cm and in depth from 4 to 12 cm. The furrow went around a large granite boulder from which two large pieces had broken off. One piece the observer was barely able to lift. The other he was just able to move, but it had evidently been thrown 5 m from the original boulder. In some places the path disappeared underground to become visible on the surface again further on. The descriptions given of the lightning phenomenon and of the track in the earth can be ascribed without difficulty to ordinary zigzag lightning, although the digging of a furrow has been reported in one other case of ball lightning.

Observations reported by reliable witnesses under well-defined conditions for viewing have convinced many that ball lightning exists, as would, no doubt, the sight of the phenomenon itself. In one unusual incident, however, the opposite conclusion was reached by the witness, J. Swithenbank, a lecturer in the Department of Fuel Technology and Chemical Engineering of the University of Sheffield. An ordinary lightning flash struck a tree approximately 100 m from a house in which there were some 25 people. The tree split, and burn marks radiated for 3 m from its base along the surface of the ground. Telephone wires to the house ran past the tree, and the telephone itself was shattered. The people in the house all saw a white ball of fire about 30 cm in diameter, which appeared simultaneously in the lounge, bathroom, and kitchen of the large house, wherever someone happened to be when the flash occurred. The ball lasted for a few seconds and to some people seemed to vanish out of a door or a window. An intense sound like an explosion was associated with the sight; the witnesses' ears rang for half an hour. Swithenbank was in a group conversing in the kitchen, and the ball appeared in the center of the group. He interpreted the event as an optical illusion because of its simultaneous appearance in several rooms of the large house and suggested that an extremely high electromagnetic pulse might have such an effect on the brain. The occurrence of phosphene, the appearance of light caused by pressure on the eye, has also been suggested as a possible source of such an illusion.

C. Collections and Reviews of Ball Lightning Observations

The primary interest and importance of observations of the extraordinary behavior of ball lightning led to repeated consideration of the rapidly increasing number of cases in an attempt to find systematically displayed properties which would provide a basis for explanation of the phenomenon. The earliest comprehensive discussion of this problem was published by Arago[16] in 1838. With additional reports collected by him as a result of the interest aroused by this work, a total of 30 observations was collected in the revision of this review for his complete works. Following Arago chapters devoted to the problem of ball lightning have appeared in several books on thunderstorms and lightning. A few lengthy reviews containing hundreds of detailed observations have appeared serially in journals. Two books devoted completely to ball lightning have been published.

Only five years after Arago's publication the leading British Electrician, Snow Harris, briefly considered ball lightning in his volume on thunderstorms[200] and suggested its formation by a type of electrical discharge which has been considered to this day. He cited the limited laboratory experiments of his time which might provide some explanation. The number of observations assembled was materially increased in 1866 by Sestier in his book on the forms and effects of lightning.[471] Flammarion, the astronomer and great popularizer of science, presented some 50 events involving ball lightning, many obtained from the meteorological literature and many collected by himself, in his widely published book[153] of 1872 on the atmosphere. Eighteen incidents of ball lightning at sea were assembled[323] in 1890. In more recent German[181] and Russian[488] works on storms and lightning the number of incidents given in full has decreased rapidly in view of the large number accumulated in the literature and the increasing use of statistical representations of the information. Brief reviews, primarily of possible sources of the formation of ball lightning, have been included in Russian,[260] Swedish,[361] and Czechoslovakian[433] volumes on lightning. The theory favored by each author is, in general, reasonable and in each case different. Observations in the Netherlands from 1916 to 1943 have been collected and reviewed with theories.[45] A book in popular style on atmospheric and geophysical events[272] published recently presents a short review which is for the most part authoritative and accurate on the inconclusive state of the ball lightning problem.

Papers presented at scientific meetings and the ensuing discussions have been published[184] in 1937 and again[108, 479] in 1965. Russian observations of ball lightning have been collected.[101] Specialized publications on ball lightning include a bibliography with abstracts of 43 papers from the period 1950 to 1960 from the Library of Congress of the United States[282] and a collection of Russian papers translated into English[434], largely on theoretical plasma structures considered for ball lightning, with an annotated bibliography containing 83 references.

The few authoritative and comprehensive sources which may be considered essential for a complete view of the problem of ball lightning contain the largest number of collected observations in addition to critical discussion of the properties which are deduced from these reports and the theories proposed to account for them. Such a source is Sauter[452] published in two parts, one covering theories, the other examples, of ball lightning, by the Real Gymnasium of Ulm in 1890 and 1892 and available in condensed form in the *Meteorologische Zeitschrift*.[453] Reports of 213 appearances of ball lightning are given in this survey, some duplicated,[65] perhaps because of the republication of reports in various journals. Sauter concluded that ball lightning occurs much more often than is commonly believed and urged scientific observers to be on the watch during storm to assist with observations in clearing up the contradictory properties ascribed to ball lightning in the available reports.

Twenty years after Sauter's work, a *Retrospective Glance at the Attempts to Explain Ball Lightning* by de Jans was published in six parts in the Belgian journal, *Ciel et Terre*.[230] This systematic review radiates the light of scientific reason and is unexcelled in citation of literature sources. Specific observations are given to provide a firm basis for the characteristics associated with ball lightning and to relate these properties to theories which have been proposed, but observations are not reproduced in full in chronological order as is common in other works of similar extent in this field. De Jans concluded that the theories then available succeeded in explaining several but not all of the properties of ball lightning. In the same period in which de Jans' review appeared (1910 to 1912) an exhaustive collection of observations assembled from the literature by Galli was published serially in the *Atti* and *Memorie* of the *Pontificia Accademia Romana dei Nuovi Lincei*.[164, 165] A later investigator[65] pointed out that Galli included in his collection instances in which the observer himself made no mention of a ball form.

As the number of observations in the literature and the number of collections increased, publication of studies from eyewitness reports presented only selected cases in full and generally replaced descriptions containing every detail each witness could recall with the reviewer's conclusions or with statistics on parameters he selected. Humphreys' study, mentioned previously, was based on 280 unpublished incidents assembled by Humphreys himself.[224] He found these reports so unconvincing that he concluded all cases of ball lightning have been a result of optical illusions or mistaken identification of other natural phenomena. McNally[321] obtained 515 reports in a survey of a large industrial nuclear laboratory. Although stationary luminous spheres which would ordinarilly be classified as St. Elmo's fire are included, a large number exhibited typical ball lightning characteristics such as motion through the air. An additional 112 observations are considered in another recent study of this type by Rayle.[420] Ball lightning cases reported in the literature were surveyed by Barry.[36] On the basis of over 400 observations, adding to those contained in previous reviews reports of the last half-century, Barry attempted particularly to distinguish other luminous phenomena such as St. Elmo's fire from ball lightning.

The best known source on ball lightning is the volume by Brand, *Der Kugelblitz,* published in 1923, in which observations are repeated fully and the properties reported are considered statistically in terms of their frequency.[65, 481] Brand found approximately 600 reports in the university libraries of Marburg, Berlin, and Gottingen and the naval observatory in Hamburg. He listed 215 of the most carefully established incidents in the hundred years preceding his publication and duplicated the original text for 108 of these. One third of the volume is devoted to consideration of the properties derived from these reports, and the theories proposed to explain the phenomenon are discussed, including particularly the limitations of the most successful theory of that time. A somewhat briefer Russian volume entirely devoted to ball lighting, Lenov's *The Enigma of Ball Lightning,* was published recently.[277] This work, popular in style but basically authoritative, lists ball lightning observations made in Russia of which only a few were previously available. Several of the more rational ball lightning theories are summarized and weighed, and the absence of a completely acceptable theory of ball lightning is noted. The difficult problems partially responsible for this state of affairs are presented.

Behavior exhibited by ball lightning in many occurrences as well as that observed uniquely in single cases presents difficult obstacles to theory. A comprehensive study of such properties is necessary for an adequate consideration of the theories which have been proposed.

Chapter 6

Photographs of Ball Lightning

Rare photographs of ball lightning are known. They provide some aid in establishing the reality of the phenomenon and in general confirm the aspects noted in observations. The photographic methods which have proven so effective with other forms of storm lightning have been unsuccessful with ball lightning, largely no doubt because of the irregular occurrence of the latter. The photographs are of two general types. One records the illuminated path evidently given by the motion of the ball. The second and rarest type is the photograph of a stationary luminous mass which, in agreement with observation, is only approximately spherical in outline and may display spark-like streamers or tentacles.

In a common procedure of lightning photography the film is exposed for extended periods of time during a storm. It is changed periodically. The film is developed and examined later to find if an interesting event has been recorded. The most valuable photographs of ball lightning, on the other hand, were obtained in snapshots; the photographer fortunately had a camera in hand and saw the object which he photographed.

The earliest pictorial records of ball lightning are, of course, hand drawn sketches by the observer. The first, published in a scientific journal[192] of 1868, is duplicated in Fig. 12.* The ball, which remained for 2 or 3 sec directly before the window through which it was viewed, was described as being mostly a bright yellow on the right, while it was largely red toward the left. Long rays of a reddish-yellow color shot toward the right becoming blinding white. Shorter rays also went to the left sloping toward the earth. Similar sketches have been provided in more recent cases[112] including some contained in the individual accounts of the previous chapter.[34, 48, 171]

The earliest photograph was published[135] in 1894, reportedly the image of lightning seen by the photographer,[230] which was for the most part round and studded with glowing spots. From it a luminous extension, a sort of train, extended to the rear, and a jagged crest extended toward the front. In the following year another photograph reputedly of ball lightning was

---

* Figure 12 is to be found opposite p. 54

published[521] with no information on the circumstances of the incident. A photograph widely published in Berlin newspapers in 1902 presumably showing ball lightning in motion was duplicated by moving the camera during a time exposure of the street lights.[593] Two additional photographs were noted briefly[312] without either description or reproduction in the accounts of the French Academy of Sciences in 1908. There were no further photographs published for two decades, and the doubts generated by this demonstration of the unreliability of the photographic image when combined with carelessness in the photographer led Brand to comment in his work that no photographs of ball lightning existed.[65]

Similar photographs of luminous traces presumably given by ball lightning in motion were reproduced[411] in a collection of photographs of nature in 1928. These were immediately classified as the result of the camera moving with its shutter open in a critique which referred to the problem of the earlier spurious photographs.[558] One of the photographs was an image of the street lamps near the Rhine formed by back-and-forth motion of the camera just as in the earlier photograph, and probable reflections in the water of the river were pointed out. Two additional photographs were attributed to the images of indoor lights and their reflections in bright traces formed by camera motion. The identity of all three photographs as true ball lightning pictures was maintained by the original author[412] in the face of this criticism. According to the photographer himself the street lights were not burning at the time the photograph was made (this does not exclude the presence of automobile headlights). In the indoor photographs, he declared there were no lights in the room, and other conditions assumed in the criticism based on spurious images were absent. The high density of the silver image, the sharp definition of the bright paths, the different widths of different paths, and the sinusoidal form of one track were cited as unquestionable evidence that the images were indeed produced by ball lightning. The formation of images of equivalent density by artificial lamps was declared impossible because of their relative faintness (compared to the presumed ball lightning). Formation of the high density silver image by a long exposure time was ruled out on grounds that the water reflections in the outdoor photograph were sharp rather than blurred, but the possibility that development time as well as exposure could affect the image density or that the water was calm was not taken into account. Another defense presented against the suggestion that motion of the camera occurred is that the tracks given by two stationary lights should be exactly parallel; the effect of a slight irregular rotational motion combined with the back-and-forth motion was evidently not appreciated. The possibility of producing such photographs with a moving camera had been generally accepted earlier and was demonstrated again in the following case.

A geophysicist was photographing a storm in Germany in 1937 through

Fig. 13. Photograph of street lights with moving camera resembling erroneous ball lightning images. [C. Bauer, *Umschau* **42**, 710 (1938).]

an open window.[302] There was a blinding flash of lightning, and the observer had the impression that a pale flame passed through his front garden. The photographer attempted to snap the picture by quickly operating his camera. The developed film showed a luminous path exhibiting loops and elbows with a second fainter trace which partially parallelled the first. A critical evaluation of this photograph, declaring that no lightning was visible in it at all, was soon published.[37] The image was declared to be that of two stationary light sources taken with the camera first in one position and then in another with some additional motion. This explanation corresponded with the photographer's later description of the actual process, although the camera was on a window sill and the photographer did not consider the movement which might be caused by the shutter release. The critical discussion was accompanied by a photograph of street lights taken to illustrate how camera motion may produce a picture typical of those presented as ball lightning photographs in which the photographer himself has not seen the ball lightning (Fig. 13). Considerable doubt is thus cast on similar photographs in which the ball lightning was not actually observed unless complete absence of motion in the camera is well established. The ultimate desirability of careful eyewitness observations in support of each lightning photograph was pointed out in an evaluation of a typical picture of ball lightning in motion, unseen by the, photographer, which resembled the image given by a moving lantern.[227,580]

The broken luminous trace falling in a smooth curve visible in Fig. 13 is of additional interest. It resembles photographs presented as images of bead lightning[454, 457] in which in which the circular or spherical aspect of the beads shown in Fig. 5 is absent. Some doubt is thus attached to such photographs[454] in which the bead lightning was not observed by the photographer. A similar photograph displaying a bright path with two portions broken regularly like dashed lines has been published[231] as the path of an evidently unobserved "thunderbolt." In another example the photographer noted some lightning phenomenon, but it is not clear that he saw the broken luminous trace in his photograph.[457] On the other hand, one observation recorded a vertical lightning in which the path was broken up into dark and light sections without specific mention of roundness in the bright beads.[2]

Examination of an original negative[295] from which a positive print evidently displaying an excellent, clear photograph of bead lightning was prepared showed that the original film emulsion had been eaten away by insects, leaving a portion of a photographic image initially of an ordinary lightning flash. One photograph reported to show bead lightning, as actually published, contained neither visible segmentation nor bead-shaped sections.[470] Fortunately, the motion picture from which Fig. 5 was obtained clearly recorded the formation of bead lightning in the channel of an initial linear flash as noted in earlier eyewitness accounts.

Fig. 6. Pinched Lightning. [B. T. Matthias and S. J. Buchsbaum, *Nature* **194**:327 (1962).]

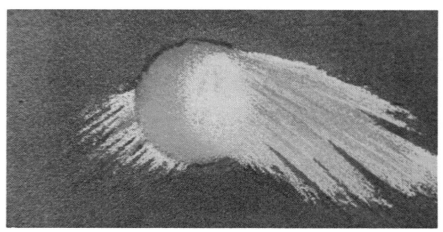

Fig. 12. Sketch of 1868 ball lightning observation. [W. von Haidinger, *Sitzber. K. Akad. Wiss., Wien, Math.-Naturwiss. Kl.* **58** (II):761 (1868).]

Fig. 14. Observed ball lightning motion. [H. Petersen, *Geophysica* **5**, 49 (1952).]

In some instances photographs depicting a luminous trace were obtained when ball lightning was observed by the photographer.[380, 418] One case of this type was considered so well established[418] that later photographs were identified as ball lightning incidents by their resemblance to it, even when there had been no direct observation.[580] Another such photograph (Fig. 14) obtained when there were two observers of the reported ball lightning[380] was declared by critics to be the result of motion of the camera when there were street lamps in the field of view because of its resemblance to previous examples of such photographs.[401, 402, 501] The eyewitnesses, however, declared that the photograph was taken during the daytime when no street lights were burning.[381] The glowing ball was seen passing the window during the film exposure. It moved slowly and passed into a cold chimney, reportedly exploding in a fireplace below. An approximate diameter of 5 cm was calculated for the globe from its distance from the camera, estimated at 10 m, and the nearly 0.5 mm width of the luminous trace on the negative. The camera was on a window sill during the exposure, and the possibility that it moved was ascribed to a push from one observer who leaned out of the window to follow the motion of the ball. In support of the authenticity of the photograph the original investigator pointed out that the discontinuity in the ball lightning trace corresponds to the displacement of the building images as expected from this camera motion.[381] In addition, light traces obtained by photographing a stationary light source while moving the camera normally show overexposure of the source compared to the trace (as in Fig. 13), and this is not the case in the photograph in question. Despite the further information from the observers, one a lecturer at the Copenhagen Polytechnic, the critics maintained that the photographs were spurious[127, 402, 467] on the basis that a daylight photograph should give a completely darkened negative in which no light trace would be visible, as well as the examples readily produced by artificial methods and the previous photographs of this type which had been shown to be false. A similar photograph presumably showing the bright path of ball lightning moving at a distance has been published (Fig. 15).[123, 277] This picture was obtained by the usual method of long film exposure with an open camera shutter during a thunderstorm. The photographer was in a different room than the camera when he noted a lightning flash and heard a crackling noise. He then closed the shutter. He reported that characteristic zigzag lines of ordinary lightning appeared on the film along with some curved lines which the photographer ascribed to an

Fig. 15. Trace of presumed ball lightning path. [B. Davidov, *Priroda* **47**, No.1, 96 (1958).]

occurrence of ball lightning. The portions showing the zigzag flash evidently do not appear in Fig. 15. The curved trace of light considered as ball lightning ended at a window on the fourth floor of a nearby honse. Examination showed a 3.5–4 cm length of the wooden window frame was charred, and on the glass were 1–1.5 cm radially diverging streaks of soot and easily broken putty. The photographer associated the crackling sound with the ball lightning.

The lightning authority, Stekolnikov, commented with due caution on the difficulty of concluding either from the report or the photograph exactly what phenomenon produced the photographic image.[489]

A complex trace attributed to a close-to-ground discharge, such as that involved in Fig. 4, was photographed at close range in a region where no artificial lights were visible.[350, 351] The photographer was observing a violent thunderstorm through an open window in 1955 in Switzerland. A flame appeared immediately in front of him simultaneous with a lightning stroke. The unexpected flash startled the photographer who suddenly moved the camera. A very complex luminous trace with multiple intertwined loops was obtained. In a detailed analysis of this trace[351] three groups of discharges were declared responsible, and the intervals between the discharges were estimated at 100 m/sec between the first and second discharges and 40 or 50 msec between the second and third. A similar discussion was published regarding photographs obtained in 1908 showing complex curvature in the presumed lightning traces, although one was taken in a well-known procedure with a camera oscillated by hand and the other with the camera reportedly held stationary in the operator's lap.[2] The latter also showed the beaded or striated sections described previously in connection with moving camera photographs. The camera motion was again not considered, evidently on the assumption that the actual light radiation was of such short duration that there could be no effect. The photographer of the second incident, furthermore, did not report observing bead lightning. The motion of the camera was not taken into account in these discussions, however, and in view of the lack of information on the observed phenomena no conclusions are warranted.

Photographs of unusual thunderstorm discharges which do not resemble ordinary lightning are often classified as ball lightning. For example, Fig. 6 was presented in one review[281] as a probable picture of ball lightning. The photographers, on the other hand, published it[313] as "pinched lightning" and it resembles no other photograph of ball lightning. The complex discharge depicted in Fig. 4, which occurred near the ground, resembles several photographs which have been published as ball lightning events and is considered one of the latter by some investigators. The discharges involved were not classified as ball lightning by the authors, however, although in one case a discharge evidently traveled in two directions from the single round, luminous source in the photograph. The luminous darts giving the traces moved at a velocity of the order of 10^7 cm/sec, according to analysis of the film, a much higher velocity than that ordinarily associated with the motion of ball lightning. This image was obtained on film itself moving at high velocity in the camera, so that the apparent lateral motion of the discharge is distorted. The photograph of ordinary storm lightning in Fig. 2 is of particular interest with respect to this problem. The second flash of lightning from the left in this figure shows, at slightly above the midpoint between the top of the flash and ground, an intense, greatly curved trace resembling the photographs attributed to ball lightning in motion. At the top of the adjacent flash, the third from the left, there is a faint curved trail resembling the near-ground discharges just discussed. The recurring consideration of such photographs in attempts to establish the properties of ball lightning thus recalls a comment made over three-quarters of a century ago that so-called novel effects of lightning are often really effects which have been well known to lightning experts for some time.[519] The acceptance of such photographs must be tempered by an evaluation of confirmatory observations of the event or by complete photographic data.

A few photographs have been obtained showing stationary ball lightning. A well known series of these was taken by J. C. Jensen, who was photographing storm discharges during a line squall while electric field and pressure changes were recorded.[233] He noted a shapeless lavender mass which appeared shortly after a flash of lightning and seemed to float slowly downwards (Fig. 16). Jensen was busy with the apparatus and took little time for direct observation, but he noted that the fireball resembled a gigantic pyrotechnic display. Two or three globes appeared and seemed to roll along a pair of high-voltage power lines approximately 600 m from the observer. They bounced down to the ground and disappeared with an explosion. The first five photographs Jensen exposed during this incident over a period of three minutes show the fireballs. The dimensions of two balls were estimated at 8.5 m and 12.8 m from the parameters of the photographs and the known distance of the power line from the camera. The altitude of the balls was estimated at 28 m above the horizon.

Consideration of Jensen's photographs has been influenced by an unusual tale which tells of student pranksters firing Roman candles during the storm to provide the phenomenon which interested Professor Jensen so greatly.[450, 580] The origin of this story is unknown. Possibly it is based on Jensen's own comment that the fire balls resembled pyrotechnics. Recent duplication of Jensen's photographic method during actual firework displays did not give similar pictures according to an investigator who has examined the original negatives closely.[272, 450]

The possibility appears, however, that the fiery masses observed by

Fig. 16. Ball lightning resembling pyrotechnics. [J.C. Jensen, *Physics* **4**, 372 (1933).]

Jensen were formed by some process involving the ordinary flashes of lightning and the high-voltage lines, a possibility indicated in a few of the ball lightning incidents related previously. Later observations were also connected with a photograph closely resembling Jensen's.[324] A large flame-like mass appeared immediately after linear lightning flashes near high-power lines. From the photograph taken at a distance of 5000 m the diameter of the mass was estimated at 10 m. Two observers much closer to the place where the glowing ball appeared independently reported a large bright flame from a transformer, evidently the object seen in the photograph, immediately following a lightning flash which struck the power line. Such a discharge would not ordinarily be free to move in the manner associated with ball lightning. In both cases, however, the photographers observed the fireball falling toward the ground and apparently exploding. The initial height of the fireball estimated by Jensen indicates that those observed by him originated above the power lines.

Two photographs of ball lightning taken at a distance of approximately 200 m and particularly well confirmed by visual observation have been published. An oval ball appeared suddenly in 1933 before a photographer taking pictures of storm lightning (Fig. 17). The intensely bright ball, which

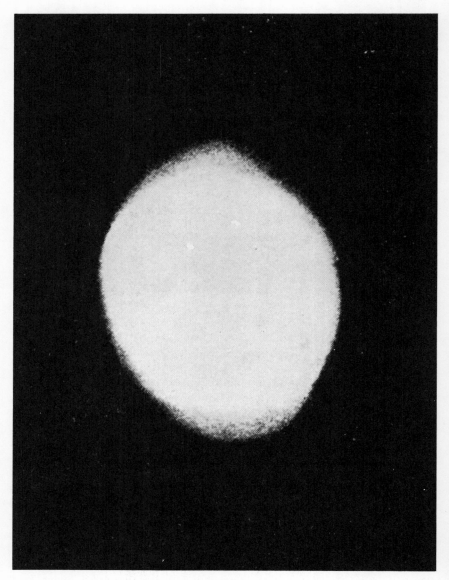

Fig. 17. Oval ball lightning. [H. Norinder, in *Problems of Atmospheric and Space Electricity,* Elsevier, Amsterdam, 1965. p 455]

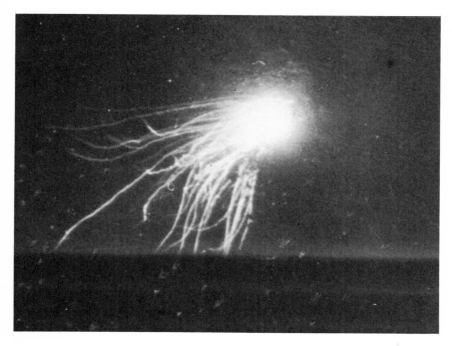

Fig. 18. Firework ball lightning. [E. Kuhn, *Naturwiss.* **38**, 518 (1951).]

was seen by several people, floated slowly toward the ground.[350, 361, 362, 459] The photographer estimated the diameter of the ball to be 35 cm, and it lasted 10 sec or possibly a few seconds longer according to some witnesses. Two minutes later a second lightning ball appeared, but the photographer had exhausted his film. A lightning ball with the typical firework appearance associated with the sketch in Fig. 12 and with Jensen's photographs has also been filmed (Fig. 18). The ball, with an estimated diameter of 50 cm, appeared to the photographer under circumstances similar to those of the previous example[248, 263]. It fell directly down and flew apart like a firework a few meters above the ground. The photographer reported that he saw the luminous body clearly at the 200-m distance and that there were no other light sources in the field of view.

The aspects of ball lightning thus recorded in rare photographs obtained as the photographer had a clear view of the phenomenon are in accord with the characteristics frequently reported in a large number of ball lightning observations.

Chapter 7

Characteristics of Ball Lightning Derived from Observations

From the continually accumulating observations of ball lightning it gradually became clear that an unusual, if not wholly contradictory, combination of properties was indicated by the eyewitness reports. An early investigator[278] summarized five major difficulties resulting from contradictory observations:

1. Ball lightning occurs both in clear skies and in pouring rain.
2. The color of ball lightning is red or blue or sometimes a combination.
3. Occasionally ball lightning is motionless; sometimes it moves very fast. Often the path is directly against the prevailing wind, or at times a faint breeze can change its direction.
4. Although the point where it originates is not usually well established, ball lightning usually travels close to the earth. It may disappear silently or explode with a bang, emitting many bright flashes.
5. Ball lightning travels on wires or along edges; or the path may be independent of such supports, the ball floating free in air or even in closed rooms.

In other words the greatest contradictions were found on all points. These disparities led another investigator to remark that many natural phenomena are not well understood, but there are very few which observation makes only more difficult to explain.[453]

Brand[65] summarized the properties displayed by ball lightning according to his extensive literature survey of observations:

1. Ball lightning is a rare, long-lasting electric discharge in ball form (less often in pear shape, among others) which is relatively more frequent in winter storms and occurs at the end of the storm. Its effects are weaker than those of linear lightning.

2. Ball lightning usually appears as a red, glowing ball or hollow ball of

10–20 cm diameter surrounded by a blue contrasting area and with vague borders. It can occasionally be blindingly white and sharply defined.

3. A hissing, buzzing, or fluttering noise can be heard.

4. After it disappears there is often left behind a sharp-smelling mist which appears brown in transmitted light, blue in reflected light, or white in moist air.

5. Its lifetime varies from the shortest fraction of a second up to several minutes; most often it lasts 3–5 sec.

6. Ball lightning may appear from the bottom of a cloud or floating free in the air or attached to some object. Often it is directly preceded by an initial lightning, and the ball originates at the place struck or a small distance from it. The preliminary lightning may be absent in more than a few cases.

7. The lightning ball disappears quietly, or with a mild bang, or with a terrific explosion in which short linear flashes shoot out in great number on all sides. Sometimes it ends when a linear lightning discharge strikes the ball.

8. The speed of ball lightning which appears from the bottom of clouds and falls down to earth is considerable (conversion to linear lightning). Near the ground or in enclosed rooms it moves at about 2 m/sec. The globes can remain entirely motionless at times; the attached ones especially can be stationary and then disappear, boiling and spraying out sparks (conversion to St. Elmo's fire). Occasionally motion caused by the wind is noted, but the path is most often unaffected by wind.

9. Sometimes several lightning balls appear around the place struck by a linear discharge. A single large ball can burst and several small ones shoot out. In rare cases two fireballs occur one over the other, connected by a chain of small luminous beads, or a short beaded chain appears with a single lightning ball (transformation to true bead lightning).

10. The free floating and attached ball lightning seem to behave entirely differently, but they can change into one another. The floating type gives the impression of very high-voltage discharges of low current, comparable to Tesla discharges. The attached ball lightning seems to be of lower voltage but greater current.

11. The floating balls have the red color of the light given by meteorites in nitrogen. They avoid good conductors and usually select a path in air. They appear to be drawn into enclosed spaces from the air, entering through an open window or door or even a small crack but especially favoring the better conducting combustion gas of the chimney, so that they very often come into the kitchen from the fireplace. After repeatedly circling the room a lightning ball may leave again, often by the same path, sometimes by a new one. Ball lightning is harmless even when it falls in a group of people since it keeps its distance from human bodies as from good conductors. Sometimes it makes two or three vertical movements of several centimeters or even meters which when combined with translational motion appears like jumping. Often the path is a single descent from the cloud to a few meters from the ground followed by a rise.

12. The attached lightning balls are blindingly bright, either white or blue. They remain on good conductors, especially right on the highest point or roll along such objects (e.g., raingutters). They heat up the substances they touch, including human bodies, on which severe burns may be inflicted when contacted, sometimes in passing under the clothing, often with fatal effect.

13. The change of a floating ball lightning to an attached one occurs after a short pause when the ball suddenly darts to a nearby conductor, particularly water. On contact with the conductor it may disappear quietly, with an explosion, or remain as a stationary ball lightning. Those falling from clouds usually continue until they hit the ground and then explode.

14. The conversion of an attached ball into a free-floating one takes place simply by its rising, usually followed by a sloping flight to the clouds, but such a ball disappears soon after, as a rule.

Brand described in this way the widely different types of behavior reported in observations and provided one of the most complete and authoritative short summaries of the observed characteristics. The need for more systematic data on the properties in question led to consideration of questionnaires, several of which were proposed and published in attempts to guide future observers. The following list presents typical questions and properties

QUESTIONNAIRE

1. Size.
2. Shape. Information in this category includes protrusions, rays, and halos or corona.
3. Color.
4. Duration.
5. Evidence of heat.
6. Motion. Information on velocity, the path, rotation, and the direction with respect to the wind are included in this category.
7. Smell.
8. Sound.
9. Emission of sparks or lightning from the ball.
10. Disappearance of the ball. Explosive or silent.
11. Traces left by the ball. Burns, damage, etc.
12. Change in appearance of the ball. Change in size or color.
13. The time of day of the occurrence.
14. Occurrence during storm.
15. Connection with flashes of linear lightning.

considered of major importance from those published previously (cf. Refs. 65, 181, 230, 321, 420, 453, 516).

Additional parameters which are omitted in the above list have been contained in the various questionnaires. Some investigators consider topographic data on the region in which ball lightning occurred significant in formation or behavior of the ball.[420, 453] The light intensity radiated and its distribution over the ball's surface, as well as whether the borders are sharply or indistinctly defined, has been emphasized by some.[65, 420] Magnetic and electrical effects from the ball lightning have also been specifically sought.[65, 321] The newer collections of ball lightning observations have been assembled by means of responses primarily made to questionnaires adapted to statistical or computer analysis.[321, 420] This method is more convenient and systematic than relying on the witness's account alone to record all the significant properties which were observed. The effect of the questionnaire, however, must be considered. The two surveys made by direct response to multiple choice questions contained no question on the direction of motion of the ball lightning with respect to the wind. This problem has been considered by some investigators as one of the major indications that many theories provide an inadequate model of the phenomenon.

The compilations of Brand,[65] McNally,[321] and Rayle[420] contain reports of over 800 observations which combined with individual reports, especially those published since 1919 (the latest cases contained in Brand), provide a large body of data on the properties of ball lightning. Brand's survey contains 215 ball lightning reports, selected from more than 600, covering a wide geographic area, although largely European, during the century preceding its publication. The two later surveys were based on the observations of employees of two technological centers. The reliability of the accounts was not further considered. In one collection[321] only 45% of the glowing masses were described as airborne or airborne part-time. Thus the remainder, which were in the majority, were not airborne at any time and were not properly classified as ball lightning in the usual sense. The review by Barry[36] summarizes conclusions derived from over 400 cases in the literature, including many in Brand's monograph and containing in addition later reports up to 1966. In the following discussion the properties of ball lightning are considered in the order given in the questionnaire above.

The size of ball lightning, considering in general a diameter estimated for an irregular sphere, has been reported from pea-size up to 12.8 m. Extreme sizes of 27 m and 260 m have also been reported.[279, 42] The balls viewed from closer distance are usually associated with smaller diameters; the larger dimensions have been reported for distant sightings in which the estimation of the size is dependent on the distance of the object, which could itself only be approximated. Brand concluded that the average size of the balls for which good estimates were available was approximately 20 cm. The diameter of the largest number seen at close distance in his collection may be as low as 15 cm. The most frequent diameter in 156 observations collected by Norinder[361] was 25 cm and in Barry's summary[36] of 400 reports 30 cm, in approximate agreement with the results of McNally's survey.[321] A somewhat larger mean diameter of approximately 35 cm is given by Rayle.[420] The distributions of ball lightning diameters from the four major collections considering this parameter are shown in Fig. 19. Approximate smoothed curves are given for each set of distributions. The data have already been averaged by each investigator by the method of reporting total number of cases in a selected interval of values for the diameter. The interval used was different in each survey. Its value is shown by the distance between the data points on each curve in Fig. 19. The approximate logarithmic distribution of the diameter of ball lightning with the total number of balls of the specified diameter or less obtained from the two most recent surveys[321, 420] has been used as a basis for suggesting the possibility of a relationship with other storm parameters which exhibit logarithmic distributions, such as the intensity of the electric field of sferics and the charge, current, and current-rise rate of lightning strokes.[420]

The form of the glowing masses is reported as generally spherical or ball-

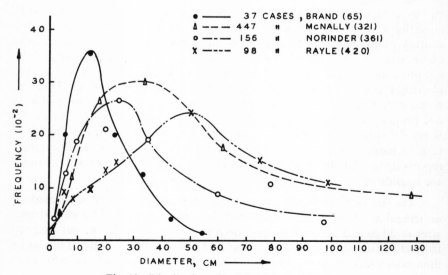

Fig. 19. Distribution of ball lightning diameters.

Characteristics of Ball Lightning from Observations

shaped; this is indicated in 83% of the cases in Brand and 87% of those in Rayle. Several of these appear hollow. A few oval or eggshaped masses have also been observed. The torus has been reported more frequently in recent observations; 62 luminous ring-shaped forms were contained in McNally's summary. Individual reports describing varied forms such as almond-, heart-, rod-, and pear-shaped are available. An irregular form is indicated in several reports of flame-like shapes. A halo or corona extending from the mass of the ball itself has been noted in several instances.[88, 259, 424]

Red and orange colors are reported most frequently for ball lightning according to the five major surveys which considered this parameter (cf. Refs. 36, 65, 164, 321, 420). Red was by far the most common color noted in the collections of Brand and Galli and the most frequent in McNally's, but Rayle found orange reported often and red seldom. Red or red-yellow were observed in 60% of the incidents reviewed by Barry. Yellow, white, blue, and blue-white (in approximate order of decreasing frequency) are also commonly reported, altogether in over 200 cases. While noting the high number of blue spheres given in other surveys, Barry found less than 2% were blue or blue-white in his study and concluded that these were cases of the corona discharge, St. Elmo's fire, erroneously reported as ball lightning.[36] He included purple or violet, however, as usually associated with ball lightning in the rod shape. Green is noted relatively rarely, and unusual black globes which do not really radiate light have been mentioned.[164, 310] Multiple colors and mixtures of colors have also been observed.

The largest number of ball lightning observations indicate that the most common lifetime is from 1 to 5 sec. An appreciable number disappear in less than a second, and the examples with a lifetime longer than 5 sec are markedly fewer. Several exhibit a lifetime of the order of 1 minute, and individual observations for 9 minutes[230, 258] and 15 minutes[65] have been recorded. Estimates of the lifetime of ball lightning often meet with the difficulty that either the origin or the demise of the ball is not observed because of its motion, in addition to the uncertainties connected with the judgment of time by witnesses unexpectedly exposed to such a startling phenomenon. The longer lifetimes, extending to periods of a minute, were correlated with motionless blue or blue-white globes in the survey by Barry, who concluded that such globes were actually St. Elmo's fire.[36]

The absence of any heat radiating from ball lightning has been especially noted as unusual for a body emitting such intense light. This property, which would ordinarily attract notice only in observations made at very close distance, is reported in by far the larger number of cases. In addition to the individual examples cited previously only four reports in Rayle's survey, in which 55 balls were seen at less than 15 m, indicated the presence of any heat. Barry suggested that the luminous globes which do not radiate heat are incorrectly identified as ball lightning and are more likely St. Elmo's fire, evidently on the basis of the high temperature expected from an electrical discharge or related phenomenon. Examples of incidents in which such a simple misclassification is improbable and heat effects were absent are given in Chapter 5, Section B, 12. Certain observations describe definite heat effects, however, including ignition of inflammable materials and burn marks left by the globe. Brand concluded that, in general, no heat effect is exhibited by ball lightning of the type which floats free in the air, whereas the balls supported on conductors (or human bodies) do burn. High temperature was also associated with the spheres of white and blue-white color.[321] A temperature of 14,000°K was estimated for the bright yellow-white ball lightning reported in Russia in 1967 by an observer with long experience with high-temperature plasmas.[128] The glowing sphere was compared in appearance to a plasmatron discharge of this temperature. The ratio of ozone to nitrogen dioxide in gas samples collected from the trail left by the ball provided an additional temperature estimate. According to the observer this ratio decreases with increasing temperature of arc discharges in air. It is usually less than one and at temperatures of 2000–4000°K is approximately 0.9. The lowest ratios found in the gas samples were near 0.8, indicating temperatures in the fireball somewhat greater than 4000°K. This value is in accord with the radiation temperature of 4000–5000°C associated in Wien's law with the common red and red-yellow colors reported for ball lightning.[35] Even lower temperatures of 200–300°C are conceivable if the glowing spheres are considered ball-shaped cool flames produced in low concentrations of combustible gases in air.[35, 36, 280, 354] The higher temperature of 14,000°K estimated for the sphere which was compared to a plasma discharge is inconsistent with the ozone: nitrogen dioxide ratios measured in the same incident. Different ratios as high as 2.45 were reported in other gas samples. In addition it may be noted that the witness made no specific comment on heat felt from the ball although it evidently passed close to him, perhaps at a distance of 2 m, during a portion of its flight.

Brand's conclusions on the greater energy of affixed and blue–white ball lightning are contradicted by one of the most discussed observations which was the basis of a widely accepted estimate of the high energy of ball lightning. This incident described previously involved a small, red-hot ball which descended from the sky during a thunderstorm.[345] It fell into a tub of water which boiled for some minutes. C. V. Boys determined in further investigation that the tub contained about 4 gallons of water which 20 min after the ball entered was still too hot for the observer's hands.[184] The energy content of the ball[184] based on the heat required to boil this quantity of water would be approximately 1360 kcal and, with the added assumption that 4 lb of water evaporated, a total of almost 11,000 kW-sec or 8×10^6 ft-lb was estimated. This total energy, to which the assumption of the amount of water evaporated contributed approximately 40%, is identical to the incorrect estimate obtained 25 years earlier for the energy of a 50-cm-diameter sphere composed of ozone[512] which is discussed in the section dealing with chemical theories of ball lightning.

The 1967 Russian observation of ball lightning, in which the gases from the path followed by the fireball were reported, provided another estimate of the energy[128]. Nitrogen dioxide concentrations from 34.7 to 1645 μg/m^3 were observed in different samples. The formation of nitrogen dioxide by arc discharges in air was reported given by

$$[NO_2] = 6.8E$$

in which the concentration in μg/m^3 is indicated as directly proportional to the discharge energy E in joules/m^3. The greatest concentration of nitrogen dioxide reported from the ball corresponds to an energy of 240 joules/m^3 released in air.

The motion of ball lightning, while extremely varied in different cases, has been accorded much significance in indicating the nature of the phenomenon. Two categories have been distinguished, for example, the luminous globes which fall to earth from the upper atmosphere and those which travel near the ground and are formed following a lightning stroke to earth or to some other structure.[213] Two similar groups appeared in the analysis of Rayle's survey. The general paths which have been observed include direct descent from the clouds to the ground (often in an almost vertical fall), horizontal flight close to the earth with the wind or sometimes directly against the wind,[94, 125, 203, 259] upward flight, up and down motion, or rebounding from the earth as some of the reports recounted earlier have shown. A largely horizontal path was observed in 54% of the cases in Rayle's collection, and 19% exhibited vertical motion. More complex trajectories were exhibited by another 19%. In several cases, the ball traveled by rolling on the ground, occasionally along wet ground (cf. Refs. 23, 168, 295, 369, 371, 384, 398, 453, 469, 522).

In a rare example of contact of ball lightning with water, a globe was reported which fell into the sea and reappeared several times, still glowing between repeated immersions.[230, 566] Travel in passageways of limited size and through small orifices has been frequently observed. There are 15 instances in Brand's collection of ball lightning entering houses by way of the chimney. Observations of this path have been made by witnesses inside the houses who saw the bright mass entering through a fireplace and by observers who were outside, as in the case of the ball entering a cold chimney and depicted in a photograph (Fig. 14),[380, 402] which was the subject of much criticism. Passage through much smaller orifices such as keyholes has been noted[453, 471]; Rayle's survey contains several of this general type. Rotation of the ball was commonly reported by observers; Brand contains 15 such reports; McNally, 46; and Rayle, 35. Several accounts describe glowing spheres which remained stationary for some time[258]; eight cases showing no motion are reported in McNally's collection, and ten in Rayle's.

Many qualitative remarks on the velocity of ball lightning are given in observations comparing its motion, for example, to that of a man walking slowly. In individual cases estimates of the distance traveled in a given time have been made. The lower velocities indicated are approximately 1 to 2 m/sec, and values have been reported[230, 279, 452] ranging up to 30 and 240 m/sec. The higher velocities are associated, in general, with observations often made in terms of angular velocities and are, thus, as in the attempts to estimate size, greatly dependent on approximation of the distance of the observer. A group of 12 of Rayle's events were associated with velocities of greater than 30 m/sec. The very low velocity of ball lightning compared to that of ordinary lightning is clearly indicated. An unusual report of ball lightning demonstrates this distinguishing characteristic in addition to other aspects of motion which have been mentioned. Two travelers found themselves in a widespread storm in a mountainous region of Germany in 1886. A glowing ball came out from a cloud and in 5 to 7 sec dropped to approximately 1 m from the ground. The ball flew with the wind at this height directly toward the two observers, approaching so close to one that he involuntarily jumped aside to avoid being struck. The size of the ball as it left the cloud seemed to be approximately that of a child's head, but at close distance as it went by the diameter was 0.5 m. The ball was bright red with a bluish envelope the width of a hand, and it radiated an intense light so that the surroundings appeared as if electrically illuminated. The globe disappeared behind a house from which a loud explosion was heard.[424] The observation of very rapid flight of a glowing mass providing a flash resembling lightning because of its velocity has been recorded.[78]

The passage of luminous globes along power lines or similar conductors is not uncommon. In addition to the observations given earlier, 103 cases of this type of motion are noted in McNally's survey and 12 in Rayle's. The airborne spheres floating through the atmosphere, on the other hand, have repeatedly and clearly displayed paths uninfluenced by conductors. The failure of lightning rods to provide any defense against these spheres was stated several times in early literature.[18, 495] An incident indicating this lack of protection was described previously.[438] In this occurrence the wall of a subprefecture 30 m from the starting point of the reported ball lightning was struck although the building was protected by a lightning rod which later examination showed to be in good condition.

The problem of providing an alternate defense against the fireballs resulted in an early suggestion that a close network of stout copper wires would be effective.[495] The use of metal grids in this method to shield against ball lightning was described in detail by a recent Russian source.[488] The method of protection specified the closing of all windows and air intakes during a thunderstorm, although the penetration of ball lightning through glass has been reported. Grounded metal grids made of 2–2.5 mm wire with apertures not greater than 4 cm^2 are to be placed over all vents and passageways which cannot be closed. Another device used for lightning protection on high-voltage lines, the Cougnard deionizer, was held by its author to be effective against ball lightning.[139] This system of lightning protection involves additional conducting lines used to round out sharp bends in the power lines. The deionizer itself consists of two parallel 60-cm-diameter steel discs 25 cm apart. This device was intended particularly for the form of ball lightning which travels along a power line. These glowing masses are often reported as ball lightning, but they have been identified in many recent incidents as a specific form of corona discharge caused by an increased atmospheric field around the transmission line combined with the normal field given by the electric power. Such discharges are thus more properly classified as St. Elmo's fire.[36] No observation of the interaction of ball lightning with either protective device has been reported.

The appearance of ball lightning has been associated with distinctive odors by observers. Smells described as being of sulfur (cf. Refs. 28, 77, 188, 199, 212, 325, 438, 444, 447) and ozone (cf. Refs. 39, 117, 138, 228, 299, 303, 305, 370, 542, 555) are common. In a few cases the odor was compared with that of nitrogen dioxide[320, 583]; one observer concluded that the smell was identical to that of a concentrated nitrogen dioxide–air mixture (and not a dilute mixture) made up for his comparison later.[320] General odors of burning have also been reported.[304, 555] Approximately one-quarter of the globes reported in Rayle's survey were associated with a smell. Ordinary lightning flashes also produce these odors, as do electrical discharges in air. In several cases this property could thus be attributed to the ordinary lightning stroke with which the appearance of ball lightning is often associated.[36] Odors have been observed, however, with ball lightning which has traveled some distance from its origin or which was unrelated to any ordinary flash of lightning. Seven examples of ball lightning with an odor which occurred with no connection to ordinary lightning are contained in Rayle's collection. The analysis of air samples taken from the vicinity of the path taken by one ball lightning in the event given in detail previously showed the presence of both nitrogen dioxide and ozone.[128]

Various sounds are emitted by ball lightning (cf. Refs. 30, 104, 190, 216, 247, 325, 420, 425, 498, 545, 546, 571, 576). Twenty-five reports of this property of the ball are indicated in Rayle's survey although further information on what the sound was not sought. The most common sound reported is a hissing or crackling noise.[216, 415, 423] The terms in which these sounds are generally described (for example, the noise of flag fluttering in the wind[423]) indicate a strong resemblance[65] to sounds emitted by an electrical discharge. In some observations ball lightning is reported as entirely silent. Barry concluded from his review that direct reports seldom mention the hissing sound, whereas this sound is definitely associated with St. Elmo's fire.[36] Observations which apparently distinctly exclude the latter are known, however, in which the witness's attention was drawn to the flaming mass by its sound.[325] The evidence of a free flight path through a window into a building, for example, indicates that ball lightning, not St. Elmo's fire was involved.

The emission of sparks or long fiery rays from ball lightning has been noted in several occurrences giving rise to a frequent description of the luminous mass as a firework.[34, 327] This aspect is depicted in some of the pictured appearances including photographs which correspond to the observer's description of this type of ball lightning (Fig. 12, Fig. 16, and Fig. 18).

The disappearance of ball lightning often occurs silently, but in many cases there is a violent explosion. A review in a work on thunder and lightning published in 1867 stated that 19 out of 20 incidents were destructive.[157] Barry's survey of the literature indicated that a majority exploded, including 80% of the red balls and 90% of the yellow.[36] Brand's collection contained 62 observations of exploding balls, and Rayle reported 24. In many of the explosions the only effect is the loud sound sometimes compared to the discharge of many cannon; in others, the ball flies into pieces, and some damage results. A Russian report of what was evidently the most damaging explosion was used to estimate the energy release from the ball lightning.[32] A mass approximately 30 cm in diameter came through the roof and ceiling of a house and escaped through a window. It traveled 50 m from the house and on touching the ground exploded, causing the house to collapse. A release of 68 kcal/cm^3 of the ball in the explosion was calculated from the surge load required to destroy a mud wall of the type involved, 190 kg-sec/m^2. This quantity of energy is seven times that given by the same volume of TNT. The total released by the ball according to this estimate was 9.62×10^5 kcal (3.0×10^9 ft-lb), or approximately 375 times that derived from the case in which a ball caused a tub of water to boil.[184]

The great energy release thus estimated for ball lightning in this case was particularly considered with respect to the protection of airplanes. Although the number of observations of ball lightning in or near airplanes in flight is very limited,[34, 183, 257, 397] explosions or damage are evidently more likely in these cases. A fireball which entered a British flying boat at 2500 m through an open cockpit window burned off the pilot's eyebrows and some hair and made a hole in his safety belt and dispatch case before entering the rear cabin where it exploded.[183] Two glowing spheres seen outside of aircraft in flight evidently caused a marked deflection in the flight path on colliding with the craft or by exploding,[34, 257] although the only permanent effect noted later was a small fused area on the tip of a propeller blade in one of these cases. Absorption of the energy of ball lightning by a cooling spray as with a Seltzer bottle was suggested as a method of protection, although some risk would be involved with very near or large globes.[309] This method using metal salts dissolved in water was proposed again 40 years later, especially for airplanes.[32]

In many ball lightning occurrences no permanent traces are found after disappearance of the ball despite its awesome activity. The damage reported in some cases can be readily traced to the characteristic results of the ordinary lightning flashes often associated with ball lightning. A total of 40 events in Rayle's survey showed effects on metal structures, buildings, and the surface of the earth, etc. The effects reported in detailed accounts typically vary from strong-smelling gases and dust raised by the ball,[320] burns in material which the ball has touched,[131] and holes bored in walls[447] to the collapse of a building caused by explosion of the fireball.[32]

No change in the appearance of ball lightning is noted during its existence for by far the larger number of cases, but in a small number definite changes have been observed in the size, shape, or color. No change was noted in 82 of the spheres contained in Rayle's survey, whereas 11 exhibited some variation. Some of the apparent changes may be ascribed to the conditions of observation, as when the ball approaches more closely to the witness in its flight. Changes in size may involve either a decrease or an increase. Brand cited three examples of shrinking and five of expanding balls, while in Rayle nine became smaller and five larger. Eight occurrences in Brand exhibited changes in shape; e.g., elongation. The light intensity of 12 cases in Rayle diminished and two increased. Color changes have also been specifically considered by Brand[65] and Mathias.[311] An attempt was made to relate the change in color of the ball from red to white to the imminence of an explosion.[311d] The same author, however, also attributed the white color to high temperature[311e] and to an iron impurity in the ball lightning.[311g] In at least one observation the change from red to white color preceding explosion of the ball lightning was reported.[49] Barry found less than 1% of the observations in his survey of the literature indicated a change in color, and all of these involved a change to bright or dazzling white of balls from the initial red, violet, or yellow colors.[36] The color change was followed by noisy disappearance of the ball. On the other hand, a change in color has been observed in only a few of the many cases involving explosions; and conversion to red occurs with equal frequency in cases cited by Brand, again in association with the explosion.

Almost half of the ball lightning observations in central Europe, considered in detail by Brand, occurred between the hours of 4 to 8 p.m. Almost 90% were seen from noon to midnight. The greatest frequency of appearance of the balls came approximately two hours later than the peak in storms during the day but otherwise roughly resembled the distribution with time of day exhibited by storms.[65] The fiery globes were most numerous in the summer months, 63% of the cases considered by Brand according to this parameter coming in this season and a total of 80% from May through September, again closely following the yearly distribution of storms. The data of Rayle's collection dealing largely with observations in the central United States also show the greatest number appearing in summer, 83%. The frequency of ball lightning is thus evidently associated with the frequency of thunderstorms although additional parameters may be involved, such as the season and time of day most favorable for observation of rare and irregular phenomena of nature under the circumstances of modern life.

The number of ball lightning appearances not directly connected with a storm is very small. Barry estimated that 90% of the cases reported occurred during thunderstorm activity. In three incidents for which reasonably complete accounts are available there appears the possibility of some distant residue of storm activity although the ball appeared under sunny skies which were clear or contained, at most, a few clouds.[140, 229, 244] These occurred after the passage of a storm but with a lapse of time, in one case, for example, two hours.[229] In another incident, a ball of fire the size of an orange reportedly fell from a clear sky onto power lines and then entered the power generating plant through a window.[82] Of the reports gathered by McNally, three indicated the formation of ball lightning under a clear sky, and Rayle reported five which did not occur in a storm. Heavy or especially violent storms took place in 79 of the incidents reported by Brand, whereas in Rayle's cases ball lightning appeared with approximately equal frequency in storms of average or violent intensity. Ball lightning or similar fiery spheres are often reported from the funnels of cyclones, tornados, hurricanes, and water spouts (cf. Refs. 90, 143, 197, 376, 442, 513, 550, 554, 594). Fireballs were noted[126] in three tornados of 11 such storms in France between 1961–1965.

The observation of ball lightning as well as other more common lightning forms during earthquakes has been noted repeatedly.[166, 504] The number of reports, including one by a staff member of the Tokyo Central Meteorological Observatory, in surveys devoted to establishing an association between the two phenomena indicates an unexplained relationship.

The majority of ball lightning incidents is further specifically associated with discharges of ordinary lightning which may appear either before or after the ball lightning. In McNally's collection of reports 378 of the spheres appeared following such a discharge, and 69 events following ordinary lightning were reported in Rayle. A significant but much smaller number, 65 in McNally and 26 in Rayle, were unrelated to other flashes. The frequency of ball lightning compared to that of flashes of ordinary lightning, considering only those of the latter which are observed to strike structures, was estimated by Brand at approximately 3%. This frequency, based on meteorological records of North German provinces from 1884 to 1899, is in agreement with data cited by Norinder[361] for Sweden from 1928 to 1938. In Rayle's survey, on the other hand, there were 180 observers of ball lightning compared to 409 who saw ordinary lightning as it struck some object indicating a much higher relative frequency of ball lightning observations, 44%.

Rayle concluded that ball lightning may be much more frequent than commonly believed, a view also presented more than a half century earlier by Sauter[453]. The relative frequency given by Rayle was extrapolated[405] to an estimate of ten million ball lightning occurrences daily over the entire earth. This conclusion led to several suggestions that the long desired systematic studies of the physical properties of ball lightning are readily accessible since, contrary to the long-held opinion that the phenomenon is rare, investigators have a favorable chance of observing it, comparable to that of encountering close flashes of ordinary lightning. Observational and recording methods designed for systematic and complete studies of the sky to capture such occurrences were proposed.[31] Actual attempts on a limited scale to capitalize on the supposed greater frequency, however, failed to encounter a single occurrence of ball lightning. The generally held opinion that ball lightning is rare was repeated by P. L. Kapitsa with the opinion that the probability of its appearance near a given observation site is negligible.[243] Swiss atmospheric physicist Karl Berger commented, indeed, that he had never observed ball lightning or found its image on photographs in 16 years of storm lightning studies, which included panoramic photography of storms.[272]

An attempt was made by Rayle to determine the correlation between pairs of 45 parameters covered in his questionnaire. No strong connection between factors such as size, duration, or brightness appeared, and the variety of responses produced no strong correlation of significant properties. The number of destructive, noisy, or very bright globes indicating high-energy content was few.

The lack of correlation among the properties of ball lightning considered was attributed by Rayle to the inclusion of a number of different phenomena in the responses to his inquiries, all designated as ball lightning. Recognition of this possibility led earlier investigators also to attempts to differentiate types of ball lightning by the point of origin, the path, or the mode of disappearance. Although such characteristics provide convenient classifications for different ball lightning appearances, no additional information on the nature of ball lightning has been gained in this way.

In contrast to earlier investigators, Barry derived from his review the conclusion that luminous spheres exhibiting certain characteristics are St. Elmo's fire rather than ball lightning. Among such properties are the blue or blue-white color, attachment to fixed grounded conductors, motion along the ground or electric lines, the hissing sound, a lifetime of minutes rather than seconds, and the absence of heat. The association of these properties with St. Elmo's fire has a reasonable basis, and the problem of ball lightning certainly is reduced in difficulty if all cases involving these properties are excluded. On the other hand, incidents in which the blue-white color was associated with high temperature and a ball floating freely in air hissed while entering a room have been mentioned. If no well established correlation was obtained among the properties considered by Rayle, one conclusion is clear from the data; the properties described in his survey are indeed those which have long been associated with ball lightning in the numerous cases recorded in the literature and, taken in sum, they are readily recognizable as the characteristics of this distinctive phenomenon.

Chapter 8

Theories and Experiments on Ball Lightning

The wide range and diverse properties exhibited by ball lightning and contained in the information gradually accumulated in the literature over the past 130 years present a difficult challenge to the natural scientist. Despite an unusual profusion of theories there is no conclusive or widely accepted explanation adequate to account for all the reported properties. The effectiveness of some theories in dealing with a limited number of the properties led in the past to hope that continued study would eventually expand the area of understanding to the more difficult characteristics. The correlation of different theories with different types of ball lightning has been considered rarely,[479] and never completely, despite the frequency with which analysis of observation has led to the conclusion that different types exist. An alternative view presented with perhaps equal frequency has been that the large number of widely different properties which elude understanding by recognized principles exactly indicate that no genuine physical phenomenon is involved. Arago responded to this argument with the query, "Where would we be if we decided to deny everything that we can't explain?"

Essential reviews indicating the changing theoretical aspects of ball lightning were presented by Weber in 1885, Sauter in 1890 and 1895, de Jans in 1910, Brand in 1924, Aniol in 1954, Silberg in 1963, and Leonov in 1965, Consideration of ball lightning theories is commonly based on classification into two major types, one in which the ball is formed and maintained during its existence by energy supplied from an external source found necessary to account for the long duration or high energy exhibited by the ball and the other dealing with globes which contain their complete store of energy and matter when formed.[230] This method of classification is not used here. The theories are discussed instead on the basis of the matter of which the ball is composed or the type of structure which is formed. A larger number of classifications is thus obtained, which may be more difficult to consider, but greater detail in comparing significant aspects of the theories with the observational data is possible.

A. Agglomeration Theories

Many of the difficult problems in accounting for the properties of ball lightning and many of the major aspects of ball lightning theories are found repeatedly from the earliest work to the present. The formation of the observed structures from the substance of which ball lightning is composed is such a problem. The earliest theory considering ball lightning as a distinctive object was presented by Muschenbroek, who is often credited with the invention of the Leyden jar.[352] He differentiated ball lightning from other fiery bodies observed in the atmosphere such as meteors. Although later investigators depicted ball lightning as a Leyden jar, Muschenbroek himself suggested that it is not an electrical phenomenon but an agglomeration of inflammable materials descending from the upper atmosphere. These materials were pictured as initially vaporizing from the interior of the earth and rising to high altitude where they condense and reassemble. Ignition or explosion finally result as the condensed mass becomes warmer and warmer as it descends toward the earth.

A similar theory to which the intervention of electricity was added was suggested over a century later.[383] Cosmic dust is impregnated with combustible gas in passing through the sun's protuberances. This substance mixed with ice crystals descends toward the earth through the hydrogen-containing upper layers of the earth's atmosphere forming small, highly electrified clouds. Near the earth the clouds start to burn like St. Elmo's fire, gradually decreasing in size. They may burn quietly because of the inert material in the mixture which retards combustion, or the production of an electric spark as the globe passes a good conductor may cause it to explode violently.

Several aspects of these theories reappear in later models of ball lightning which will be discussed. For the present, however, these historically interesting explanations may be rejected on grounds that no reason is apparent for the condensation of the widespread matter into a concentrated mass of small extent in the upper atmosphere; and the occurrence of ball lightning produced in this manner would exhibit no particular association with thunderstorms.

B. Leyden Jar Structures

The association of ball lightning appearances with storms showing great electrical activity soon led to theories which compared the structure of ball lightning to that of the Leyden jar, the earliest method by which electrical charge could be stored in a geometric form comparable to that of ball lightning. The first such suggestion was based on a physical theory of the spheroidal state of matter, presumably a specific form of matter in bodies displaying little interaction with their surroundings.[7, 392] According to this concept ball lightning is a globe formed by the condensation of the electric fluid. The sphere is surrounded by a very thin elastic film and is filled with light gases. The color observed is a result of the reflection of daylight from the thin envelope which produces interference colors.

The earliest structure of ball lightning specifically based on that of the Leyden jar[506] was suggested in 1859. A spherical layer of dry air compressed by the attraction between the two opposite charges accumulated on either side of the layer plays the role of the insulating glass of the Leyden jar. This charged bottle with gaseous walls is said to be in a stable equilibrium maintained by the balance of the radial forces: the atmospheric pressure, the force between the two opposite charges accumulated on either side of the insulating gas wall, the pressure in this wall contributed largely by the electrostatic forces, the reduced internal pressure of the ball, and the force of the charges on the internal face of the insulating wall. The luminosity of the ball is caused by slow combination of the opposite charges across the compressed-gas layer, which is not a perfect insulator. The recombination produces ozone. This structure is not destroyed on contact with the ground, which only removes the small amount of free charge on the outside surface; but if the insulating layer is pierced by a conductor, an explosion is obtained from the discharge between the opposite charges. The equilibrium of the radial forces in the ball remains stable even with slight variations, and there are many different ways in which this equilibrium and thus ball lightning can be produced.

A singular result observed in an experiment with a Leyden jar in the 18th century provided some support for this view of ball lightning, although it was initially related to the properties of bolides, at that time considered an electrical phenomenon.[46, 158] A ball of fire like a red-hot iron ball approximately 2 cm in diameter and rotating rapidly was seen in a Leyden jar as it was being charged. Suddenly there was a loud explosion accompanied by a bright flash, and the glass of the bottle was pierced by a circular hole.

The Leyden jar theories were widely accepted. Some investigators considered that the presence of both positive and negative charges accounted for the lack of attraction to conductors such as lightning rods.[495, 506] The role of dust or plaster in providing the nonconducting wall for this structure was suggested.[369] The charged sphere was also said to form by the separation of a portion of a cloud with an electrical charge, resulting in a structure composed in part of water.[7, 363]

The occurrence of two types of ball lightning was suggested, the floating Leyden jar and a true ball-shaped lightning.[93] The first, although surrounded by an insulator, is still connected with the cloud in which it originated, and the thunder heard comes from the cloud rather than the luminous globe.

The absence of a good reason for electrical charges of opposite sign to separate in a spherical form in the atmosphere rather than to recombine directly, or for a thin layer of atmospheric gases, however compressed, or water droplets, or dust particles to prevent diffusion of such charges under electrostatic forces presents considerable difficulties for this theory. The equilibrium of radial electrostatic forces given by opposite charges gathered in concentric spherical shells is unstable, and the charges would recombine in a very short time, possibly a few milliseconds in the atmosphere, according to experimental observations of the behavior of ions and electrons.

C. Transformation of Linear Lightning Into Ball Lightning

The reappearance of theories after long intervals of time, as much as a century, has not been unusual in studies of ball lightning. The view that ball lightning is composed of the substance of ordinary lightning and is formed by the separation of a portion of an ordinary flash of lightning has been repeated in this way. A close apparent relationship may have been the basis of Aristotle's discussion of types of lightning,[17] and the identification[240, 348, 375] of ball lightning as stationary lightning, of linear lightning as ball lightning in rapid motion, and of bead lightning as multiple ball lightning is not uncommon. The numerous observations of ball lightning closely following ordinary lightning provide some support for this theory.[238, 522] Examples indicating a very close relationship were published as early as 1850, when luminous spheres were seen traveling along the path of a previous linear flash and appearing at the lower tip of ordinary lightning.[103] The complete process of formation of a separate ball from the lower end of an unusual lightning flash was observed directly.[256, 361]

The direct observation of the appearance of ball lightning from linear lightning is strong support for this theory; but several difficult questions remain unanswered in a theory of this general form, and the formation of ball lightning in closed rooms, for example, cannot be accounted for by this mechanism. The formation of a fireball inside a room directly after lightning struck very near outside led to the suggestion in 1892 that the external flash produced the object inside by electric induction.[301] The observer had been standing at the window during the storm. Hardly a half-second after the linear lightning he become aware of the ball as it detonated and disintegrated into small sparks a short distance above his head.

The surface tension of the substance of the lightning channel, called "fulminating matter" presumably from its explosive characteristic, was proposed in a series of papers from 1924–1934 as the cause of the formation of ball lightning from ordinary lightning and as the basis of several ball lightning properties.[309, 310, 311] Fulminating matter was described[311] as being composed only of the gases found in air and of enormous energy, with an approximate temperature of 2500°C. The formation of ball lightning from the channel of ordinary lightning was pictured as a cooling process accompanied by increase in the surface tension leading to separation of the ball. The outer layer of the ball is a cool elastic fluid; the interior is incandescent. The stability of the sphere, as in cases reported of ball lightning bouncing, was ascribed to the surface tension. The division of a single fireball into several smaller ones could be a result of decrease in surface tension from absorption of impurities by the hot globe, e.g., iron from contact with metal or sulfur from vegetable and animal matter. These impurities can decrease the surface tension and thus cause disruption of a ball even when pure fulminating matter would normally increase in surface tension as its temperature falls. Other lightning forms occasionally reported, such as serpentines, were also credited to the decrease in surface tension.

The duration of the bright light from ordinary lightning is much shorter than the usual lifetime of ball lightning, with the exception of certain rare flashes including the one from which ball lightning was seen forming which seem to exist for several seconds. The conditions responsible for the rounding of the advancing end and for the much longer life of the substance of the ball, if it is the same as that in ordinary lightning, have not been explained.

D. Generation of Ball Lightning by Chemical Reactions

Arago in the earliest comprehensive discussion of ball lightning[16] suggested that it might contain compounds produced from air by the effect of lightning. This view was evidently based on the experiments of Cavendish in which nitric acid was formed by electrical discharges in air. In ball lightning the chemical substances were thought to be thoroughly mixed with lightning matter. Arago pointed out the difficult questions of how and where ball lightning is formed, of what substances, and how its variable course is determined, questions which have not been conclusively answered to this day.

The generation of nitrogen triiodide by electrical discharges in storms from atmospheric nitrogen and traces of iodine or iodine-containing compounds was an early suggestion made primarily to account for the fulminating or explosive character of ball lightning.[47]

The compounds formed from nitrogen by oxidation, especially nitrogen dioxide, have been more frequently suggested in chemical theories of ball lightning. The direct comparison of the odor of a gas in a ball lightning appearance with nitrogen dioxide[320] and the formation of nitrogen dioxide by electrical discharges in air[113] provided some support for this theory which follows Arago's original discussion. In rather poorly supported efforts to explain with this view some of the more difficult aspects of ball lightning, the violent explosions often reported were ascribed to immediate nitration of organic material, such as hay stacks or trees, by the concentrated nitrogen oxides.[466] In one discussion a degree of stability up to 3000°C was attributed to nitrogen dioxide.[184] The rate of nitration indicated woule be much greater than that usually observed, particularly with a dense solid such as a tree, and explosions of ball lightning in midair and on very brief contact with substances which were not or could not be nitrated are not in accord with this explanation. It also appears[372] that nitrogen dioxide would be almost completely dissociated at 1000°C. The exothermic reaction of nitrous oxide with hydrogen forming nitrogen and water has been suggested.[100] The estimated of high energy content of ball lightning obtained from its effects in an explosion[32] and in the heating of a tub of water[184] were held to exclude the presence of the relatively low-energy nitrogen oxides. Measurement of the composition of the air following a very long and intense thunderstorm indicated no increase in the nitrogen oxides but a large rise in the concentration of ozone,[428] although the failure of ball lightning to appear in this storm prevents a definite conclusion on its possible effect.

Ozone has been favored over nitrogen oxides in some theories. Factors cited favoring ozone as the principal substance in ball lightning were the odor observed on disappearance of the ball, the density greater than air displayed by the descending spheres, the deflection of the falling path near the earth followed by horizontal motion (said to result from the negative charge of both the earth and ozone), and the blue color observed in electrical production of ozone in contrast to the usual yellow nitrogen flame.[512] Ozone was reported to form near highly charged points, particularly those of negative charge. Nitrogen oxides were said to be absent, appearing only when sparks stream from the points. The unusually high density of gases in the ball deduced from its rapid descent and supposed high temperature above 1000°C led to a suggestion that polyatomic oxygen and nitrogen molecules up to O_{12} and N_{12} are involved.[309] Consideration of polyatomic carbon molecules (up to 17 atoms) in high-temperature (ca. 3500°K) globes containing seven-eighths carbon vapor and one-eighth air indicated that the density would be below that of air under ordinary conditions.[293] The energy released by a sphere of ozone 50 cm in diameter was estimated at 11,000 kW-sec (8×10^6 ft-lb), the identical quantity also derived later by calculations based on the observation of a tub of water heated to boiling by a lightning ball.[184]

Recalculation of this quantity with recent values of the heat of formation of ozone[372] for such a sphere of pure ozone reverting to oxygen gives only approximately one twenty-fifth of the energy indicated. The points presented in support of the presence of ozone, however, do not rule out nitrogen dioxide; for in at least one report the odor of nitrogen dioxide was identified in gas of an appropriate orange color associated with the appearance of a fireball inside a house.[320] The odor was compared with a somewhat concentrated nitrogen dioxide–air mixture prepared later for one of the witnesses. Furthermore, the density of nitrogen dioxide is almost as great as that of ozone, and yellow or red colors have been reported in many more cases of ball lightning than blue.

The presence of both ozone and nitrogen dioxide as gases produced by ball lightning (rather than as substances composing the fireball or primarily responsible for its properties) was reported in the Russian observation of 1967 described earlier.[128] Explosion of ball lightning, which did not occur in this case, was attributed to chain oxidation of nitrogen caused by the reaction of ions. Samples of gas were collected in evacuated bulbs held close to the path followed by the fireball approximately 1 min after its flight. The nitrogen dioxide was determined with a mass spectrometer and ozone by its reaction with tritium-containing sorbents present in the bulbs. The other gases in air were found in their normal concentrations. These two were found in highest concentration in the sample taken very near the path of the ball which contained 52 times the normal ozone concentration and 110 times the normal concentration of nitrogen dioxide in air. The ratio of ozone to nitrogen dioxide varied from 0.78 to 2.45 in the different samples, as shown in Fig. 9. The ratio was highest in the first sample of gas collected and lowest in the second, which contained the greatest concentrations of both. The formation of ozone and nitrogen dioxide by electrical discharges was investigated experimentally. The ratio of ozone to nitrogen dioxide decreased as the discharge voltage increased. The concentration ratio of 2.5 would correspond to a lightning potential of 300 to 400 kV if the natural discharge were compared to the silent discharges which were studied. A range of one to six was observed in this ratio in the experiments. An increase in the temperature of the discharge also reduced the ratio, which was usually less than one for the arc discharges. A value of approximately 0.9 was found at 2000–4000°K. The lowest ratios, near 0.8, would indicate that the temperature of the lightning ball was greater than 4000°K. The observer commented, however, that the luminosity of the ball was comparable to that of a discharge at 14,000°K.

The production of hydrogen and oxygen, gases which give explosive mixtures, by electrolysis of water provided an early basis for proposals that ordinary lightning discharge usually seen before ball lightning produces an explosive gas which is electrically charged.[209, 210, 211, 436] An aqueous wall may surround this gas, and a spark from the electric charges causes the explosion. Similar theories were suggested almost a century after the first[436] involving oxygen and hydrogen formed by thermal dissociation of water from the high temperature given by a strong flash of lightning[42] or by normal electrolysis.[461] Hydrogen was reported present in the air samples collected near the path taken by a ball lightning.[128] The average concentration was given as 1.2 mg/m^3, and the ratio of hydrogen to oxygen was 4.3×10^{-6}. This ratio is far below the lower explosion limit of these gases which occurs at a weight ratio of 0.0026 (or 0.04 volume ratio).[280] The reported concentrations do not support the theory that an explosive mixture of hydrogen and oxygen produces ball lightning. The theory is not disproved, on the other hand, since the gases analyzed were collected in and near the trajectory, not inside the fireball. Thus, the results may represent substances remaining after any chemical transformations which occurred rather than the concentrations present before passage of the ball.

The generation of luminous spheres by ignition of low concentrations of combustible substances in air succeeded in duplicating many of the properties displayed by ball lightning.[354] Hydrogen methane, propane, and benzene were ignited by brief electrical discharges in vertical cylindrical glass tubes. Hydrogen, with which the effect was obtained most readily, gave a faint blue light visible in darkened rooms at concentrations between 3.8 and 9% in air. With propane a minimum concentration of 1.24% was necessary. The minimum concentration of benzene giving the globe was too low to measure, but if a few drops were placed in the tubes and allowed to evaporate the glowing sphere could not be produced until the concentration of benzene was lowered by evacuating the chamber to low pressure several times. Propane and benzene gave brighter clouds which were easily visible, usually blue, green, or violet. The lower and upper limits of concentration in formation of the globes are analogous to concentration effects noted in the flammability limits of gases, and these luminous clouds were described as diffusion flames initiated by the electrical discharge. With concentrations of the fuel gas above the limit ordinary combustion flames might form as with propane in air.

The masses formed under proper conditions resembled a corona discharge, a hollow mass glowing on the surface, rather than a burning gas. Spherical, drop-shaped, and pear-shaped forms were produced which rose through the cylinder at velocities between 0.3 and 10 m/sec. These clouds exhibited several additional characteristics reported in observations of ball lightning. Although usually noiseless they occasionally emitted a hissing sound when generated with higher concentrations of the fuel gas. In a few experiments there was a loud report when the globe reached the upper cover of the experimental vessel and disappeared. Magnetic or electric fields did not affect the motion. Only moderate warmth was emitted by these bodies.

Occasionally a very sharp and choking odor remained in the experimental chamber, which was attributed by the investigator to formation of nitric oxide by the spark discharge. The partial oxidation of hydrocarbon gases under these moderate conditions may give such an odor as well as the light brown fog observed by the formation of peroxidic products, resembling the effects of atmospheric processes on such compounds from automobile exhausts. The separation of a larger ball into small parts which reunited was observed, and in a few experiments the oxidizing mass passed through a 7-mm hole in a brass disc obstructing its path and emerged on the other side with its normal dimensions.

Later studies reproduced the formation of the glowing spheres on ignition of hydrocarbon–air mixtures at concentrations below the combustion limit.[35, 150] With propane as the fuel an ignition spark gave no visible result below the combustion limit (ca. 2.8%) until the lower range of 1.4 to 1.8% in air was reached. The spark released approximately 250 joules in 1 msec. At the lower concentrations a yellow-green fireball a few centimeters in diameter appeared. The sphere was rather bright and moved rapidly around the chamber for approximately 2 sec until it disappeared silently.

Two additional questions concerning this fireball were considered: how the fuel concentrations needed are formed in nature and how the size of the globe is determined by the hydrocarbon–air mixture. The transformation of the simple hydrocarbons initially present to complex organic molecules by an atmospheric discharge of electricity[396] was suggested. Further clustering of the complex molecules into large charged particles[92] similar to an aerosol was proposed in the field of the atmospheric discharge presumed to ignite the mixture. Processes for concentrating the small natural abundance of methane distributed in the atmosphere (approximately 10^{-4}%) are needed if this explanation is to apply in other than marsh regions where the fuel gas might normally be in sufficient concentration. An increase in the concentration of fuel gas to the normal combustion limit was evidently considered necessary. The occurrence of combustion in a spherical flame permits use of the relation for a normal spherical flame to estimate[280] the size of the fireball which would be formed from

$$H = \frac{d^3 c \rho \pi}{6}(T_b - T_a)$$

in which H is the heat added to the ball of diameter d by the combustion which raises the initial temperature of the gas T_a to the final temperature T_b. The fuel–air mixture has a specific heat capacity c and density of gas ρ at T_b. The diameter of the spheres expected can now be estimated using representative values of the parameters. The concentration of hydrocarbon is assumed below the minimum combustion limit initially, which is 5.4% (by volume) for methane in air and 2.8% for propane. Complete combustion of such mixtures in spheres of the size of interest releases 10^2 to 10^6 joules. The mixture of gases produced woule have a specific heat capacity[280] of approximately 0.28 cal/g-°C. and density of 2×10^{-4} g/cm^3. The red or red-yellow color commonly reported in ball lightning is associated with a radiation temperature of 4000–5000°C according to Wien's law. The diameter derived with these values from the spherical flame–energy relationship is in the range 6–130 cm, in accord with observations of ball lightning.[35]

Some aspects of the more detailed combustion fireball may be considered critically without diminishing the value of this model in accounting for some of the difficult properties associated with ball lightning. The formation of more complex hydrocarbons from methane by an electric discharge was shown experimentally for concentrations greater by an order of magnitude than for those considered in connection with the natural fireballs.[396] At the lower concentrations the electrical condensation may be ineffective. The agglomeration of fine particles by an electric field into large, easily seen clumps suspended in air by their charges was shown with a crystalline substance.[92] The formation of analogous small liquid aerosol droplets and their behavior in the natural field, as depicted in the ball lightning mechanism, is not necessarily the same under natural conditions. The temperature estimated for the fireballs from Wien's radiation law is double the normal flame temperature under optimum conditions[280] for either methane or propane burning in air, approximately 2000°C. At the low concentrations previously indicated, the combustion temperature would be lower than this.

The appearance of the fireball effect at very low concentrations of benzene, an already relatively complex molecule, suggests that some mode of reaction differing from normal flame combustion is occurring. A low temperature oxidation of hydrocarbons is known at concentrations below the normal combustion limit. Pentane in air, for example, undergoes oxidation when heated[280] and emits a blue glow as the "cool flame" proceeds at a temperature of approximately 220°C. The blue radiation has been attributed to formaldehyde produced in an excited state by the reaction. The concentrations of fuel gases at which the fireballs were observed[35, 354] are approximately those at which cool flames occur. The process for concentrating the hydrocarbon to obtain normal combustion would thus no longer be necessary to account for the experimental observations. On the other hand, the cool flame phenomenon has not been observed with either methane or benzene and the latter gives an easily visible fireball in this experiment. Furthermore, the formation of a bright yellow-green ball in the propane–air mixture contrasts with the faint blue radiation normally associated with cool flames. The radiation of molecular oxygen ion species, O_2^+, occurring at 5586Å and 5632 Å is a possible source of the reported color. The formation of these ions presents such a high energy requirement (12.5 eV or 288 kcal/mole) that excitation by the continuing fuel–air reaction of these energy levels in ions produced initially by the electric spark is possible, at best, in the absence of a specific chemi-ionization reaction. Over half of the heat of complete combustion of propane would be required in formation of the ions from oxygen. Other slow combustion–oxidation reactions may proceed at fuel concentrations even below those found in cool flames.

The spherical structure of the fireball has not been fully considered. The globe is small, constant in size, and in continuous motion at a distance from its starting point. The usual spherical flame might show only radial progression from its ignition point and possibly upward motion.

The release of marsh gas (methane) or alternatively hydrogen and oxygen, active nitrogen, or nitrous oxide for the occurrence of this process in nature was suggested to result from the heat of a large current lightning flash to ground at mountain streams.[460, 461] According to this theory ball lightning is a secondary phenomenon initiated by ordinary lightning and possibly maintained by the reaction of chemical compounds also produced by the previous flash. This view has been accepted by some authorities.[433, 488] Certainly the appearance and many of the characteristics reported in a large number of ball lightning observations were duplicated in the experiments on electrical ignition of low concentrations of fuel gases in air. A number of limitations appear, on the other hand, with attempts to extend this theory to additional properties of the natural phenomenon. Luminous clouds formed in this manner would be readily set in motion if not completely dissipated by slight air currents. The flight of ball lightning with prevailing winds noted in several reports is in accord with this expectation, but the path of others directly against the wind is not. The larger number of experimental globes rose through the vertical chambers in which they were prepared, as would be expected from warm gases particularly those most likely to be found in nature, hydrogen and methane, which are lighter than air. An unspecified number of the experiments were made in horizontal cylindrical tubes, and presumably the glowing masses were able to traverse these tubes. The method of ignition with a pair of pointed electrodes near one end of the vessel may readily give a directional force. While the fuel gases required for this method of formation of ball lightning might occur in several locations in nature, particularly, for example, over marshes, the generation of the fiery masses indoors (if not near artificial gas fixtures) is not readily explained. Balls formed close to earth following a flash of lightning or in electric fields sufficient to give corona discharges and displaying, in general, moderate temperature, low energy, and a rising path can be accounted for in this theory. The formation of the reactive gas mixture in a localized region and its ignition by the initial flash of lightning striking a specific area provides a reasonable process lacking in other chemical theories by which the required substances may be obtained in a volume of relatively limited extent.

E. Nuclear Theories

The appearance in storms of high electric fields with potential differences capable of imparting to charged particles, such as alpha particles, the velocities observed in radioactive disintegration led to one proposal that even greater localized storm potential differences may cause nuclear reactions in atoms of the atmospheric gases.[61] According to this theory ball lightning would exhibit radioactivity. The production of neutrons in thermonuclear experiments was followed by the suggestion that thermal neutrons are produced by lightning discharges, and these neutrons react with atmospheric nitrogen to produce the radioactive carbon 14 isotope of which ball lightning is formed.[121] Protons generated from water in the air by a lightning discharge and accelerated to 1 MeV by storm potentials were then proposed for nuclear reactions.[7a] The radioactive oxygen and fluorine isotopes, ^{15}O and ^{17}F, would be formed from ordinary nitrogen and oxygen in the air, ^{14}N and ^{16}O. The short half-lives of the products, 124 sec and 66 sec, respectively, would give an intensely radioactive fireball compared to ^{14}C with a half-life of 5730 years. The properties of ball lightning were attributed to the energetic positrons emitted by the short-lived isotopes. Measurement of the abundance of the decay products, ^{15}N and ^{17}O, after tornados, intense thunderstorms, or ball lightning was suggested as a test of the theory.

The focusing of heavier reactive particles, such as cosmic rays, by thunderstorm fields into ball lightning, which involves a self-sustaining nuclear reaction was also theorized.[15, 97, 277] The accompanying conclusions would be, however, that solar flares, which increase the cosmic ray flux markedly, should increase the number of ball lightning cases and the fireballs should occur largely at high altitude.[277]

Energy loss to the atmosphere would, among other obstacles, prevent acceleration of charged particles in the atmosphere by storm potentials to sufficient velocities for nuclear reaction. The peak temperature observed in the lightning discharge is too low by orders of magnitude to provide the plasma conditions suitable for thermonuclear reaction. Although cosmic rays reach the earth with great enough energy for nuclear reactions, they are too diffuse to provide energy in the concentrated regions occupied by the fireballs, and the focusing of such particles as suggested requires strong fields adequate for the high energies of the particles, formed in a lens of vast extent.

F. Charged Dust and Droplet Models

The difficulty of explaining the long continuing luminosity of ball lightning resulted in one suggestion which compared this light to the radiation from biochemical processes such as fermentation.[500] Atmospheric pollen and dust provided substances which cause the light-emitting reactions according to this theory. The idea of electrical charging of droplets and dust particles in clouds to form ball lightning was presented much earlier, in 1855, when ball lightning was considered an example of the favored spheroidal state of matter.[392] Small electrical sparks, spherical and blue-red in color, were produced between plates of insulating materials such as rubber or glass in the presence of water or dust.[278] The small size (approximately 0.5 mm in diameter) and the unstable, readily influenced motion of these sparks provided only weak support for charged particle theories. The additional discrepancies as in the failure of the natural spheres to disappear on contact with conductors as these experimental balls do and the question of how such a discharge could travel into dwellings were pointed out by the investigator himself.

The appearance of fine solid particles has been frequently associated with ball lightning.[233] One investigator ascribed a majority of the fireballs to dust activated by storms, especially volcanic dust clouds.[465] The Russian physicist Frenkel suggested that lightning generates reactive substances in the atmosphere which condense as droplets on dust in the air.[162, 277, 278] The dust particles are heated to glowing by the lightning discharge. The condensed activated gases form a thin envelope giving a structure comparable to a bubble. This model resembles the Leyden jar structures described earlier. Frenkel later withdrew the hypothesis of the spherical film and described the ball shape as a vortex formed by the mixture of charged smoke and chemically reactive gases. A similar whirling assembly of charged hail, graupel (sleet), or street dust was suggested to explain ball lightning in motion.[440] The familiar entry of the glowing balls by way of the fireplace was related to the dust theory by the process in which an initial linear flash of lightning discharges in the chimney, charging the ashes and producing a small vortex.[465]

An aerosol of charged solid particles was produced by distributing a fine dust with a fan inside a glass chamber containing one electrode with an electrostatic charge.[92] When circulation by the fan was ended, an almost spherical concentration of the particles 20 cm in diameter gradually formed in the center of the chamber. The insertion of a charged wire into the chamber repelled or attracted the sphere depending on the sign of the charge on the wire, and the sphere returned to the center of the chamber when the wire was removed. Use of a dye such as p-xylene-azo-β-naphthol produced a bright red sphere, although when finely dispersed in a Tyndall beam the particles were greenish-blue even after charging. The assembly formed in the chamber, however, contains larger particles which are red. Microscopic examination of some of these showed they were in the form of long ropes or chains, which are often observed on charging of small particles at electrodes. Formation of the sphere was ascribed to repulsion of the particles by the wall of the glass chamber, both bearing the same charge induced by the electrode. The generation of ball lightning by a similar process was suggested.

Consideration of an observation in Sauter's collection in which 25 to 30 blue balls rolled down a path during a heavy rainstorm[452, 453] encouraged reinvestigation of the electric charging of water for the theory that ball lightning is a charged bubble.[557] A German forester overtaken by a storm in 1874 witnessed the balls from the shelter of an old cabin directly on the path. While lightning flashed all about, the globes, which were approximately the size of nine-pin bowling balls, rolled rapidly in the same direction, often one right after another, and disappeared near him with a crackling sound and the ejection of blinding sparks but without exploding. The luminous balls in this incident were described as charged water bubbles rather than lightning.[557] Their formation was attributed to a high ground charge caused by the storm transferred to bubbles of the falling rain, the motion of which is then affected by wind and by electric fields. Destruction of the bubbles can result from evaporation of the water or by neutralization of the ground charge by a close flash of lightning. Formation of a charge on buildings by storm activity was suggested to explain the motion of the charged bubbles into houses.

Experimental studies of this theory were made by forming water bubbles on an insulated tube surrounding a charged wire. A metal disk was used as the other electrode at 50 cm from the tube. A bubble was formed on the end of the tube, and the electrodes were then charged. The bubbles were torn off the tube and flew directly to the disk, breaking on it with an easily heard pop. No light could be seen from these bubbles even in a darkened room.[557] Additional studies were made[354] with an arrangement more closely simulating the natural conditions, water-covered ground in an electric field, involved in the theory. Gravel and sand on a metal plate were covered with water through which air was bubbled. A potential was applied between the metal plate under the materials, evidently resembling rain-soaked ground, and a plate some distance above this surface. The field caused a marked change in the motion of the bubbles which initially were rising slowly. Their paths varied greatly, some rising with different velocities, some remaining motionless between the plates, and others reversing their paths. No radiation was observed, and perhaps because the bubbles did not assemble in a recognizable form the investigator concluded that there was no resemblance to ball lightning. The appearance of ball lightning in storms with no rain and in other cases by flashover on high-voltage lines were also cited in opposing this theory.[354]

A later theory of the structure of ball lightning was based on positively and negatively charged water droplets in a sphere, all the particles of one charge forming a ball in the center of the sphere surrounded by a shell of the droplets with the opposite charge.[416, 417] The rate of neutralization by recombination of the opposite charges in this model is moderated by the vaporization of water by the heat released. The water vapor thus formed provides an insulating wall between the regions containing the opposite charges. This structure resembles closely the Leyden jar theories previously discussed. The reason for formation of the spherical-shell structure with separated charges and the long existence of the balls in several reports compared to the short time in which recombination would occur remain as unresolved questions in this theory.

Luminosity has been observed from water drops of opposite charges falling through air.[331] Drops of approximately 500-μ radius bearing a charge of 5×10^{-11} coulomb produced light faintly visible to the dark-adapted eye when there were 400 droplet collisions per second. The light was actually emitted in intense pulses which appeared faint because of their short duration. Molecular nitrogen species were evidently responsible for the radiation, which fell in the range 3371–4285 Å, the violet and blue region of the spectrum. This radiation was not exhibited by droplets falling through carbon dioxide. Breakdown of the gas in the fields caused by the droplets was held responsible for the light. Luminescence of water was previously observed on application of high sonic frequencies and ascribed to the mechanical agitation at the gas–liquid surfaces of bubbles formed in the liquid by dissolved gases. Potentials sufficient to give breakdown were present in the experiments with charged drops although recombination of charges can take place without breakdown if the drop parameters change. These experiments[557] provide some support for the suggestion of the original investigator that light might be produced by charged aggregates of water formed naturally in the greater fields of storms, although none was observed in the early laboratory experiments. The extremely faint luminosity reported from interaction of oppositely charged water drops would presumably increase in intensity with more collisions between drops over an extended period of time.

The discharge of electrostatic potentials on rain droplets has been described as the source of the short radio waves assumed in recent theories for generating ball lightning.[8, 9, 291] This process is discussed in a later section on natural electromagnetic radiation theories of ball lightning.

The major role given to water drops and ice particles in theories of storm electrification has been paralleled in a number of ball lightning theories from earliest consideration of the problem to the present. Models such as those discussed thus far, however, have failed to show that sufficient concentrations of these charged particles can be generated and confined in nature in the forms exhibited by ball lightning.

G. Molecular Ion Clouds

Immediately following the experimental studies with charged solid particles in 1931 the suggestion was made that ball lightning is composed of molecular ions of gases rather than the much larger dust particles.[430] The interior of the ball was viewed as an attenuated charged gas. The whole was compared to an evacuated light bulb in explaining the evidently violent explosions which occurred indoors without damage. Ion theories of ball lightning deal with substances closely similar to those in plasma theories although the theoretical methods are quite different. Later ion theories presented, as opposed to plasma structures, attempt to distinguish the material in the ball from a plasma by its low temperature, low charge density, or the absence of free electrons.[213] Thus, the theory of generation of ball lightning by an ordinary linear flash forming a substance composed of positive ions and free electrons at the same density as a gas at normal pressure and temperature[338, 339] would now be recognized as involving a plasma which can hardly be produced even in thermonuclear laboratory experiments. An ionized gas is commonly considered a plasma if the distance over which the electric field of a selected charged particle extends, as given, for example, by the Debye shielding distance,[285] is small compared to the dimensions of the ball. The plasma is neutral when investigated by methods which can detect charge only over longer distances than this. It appears that the plasma description can be readily extended in general terms, including that of neutrality, to the molecular ionic gases (occasionally supposed to be distinctive substances) with the benefit of progressing beyond qualitative descriptions.

A sphere consisting of a layer of ozone or nitrogen ions from lightning or a brush discharge was suggested for ball lighting.[374] As in the earliest structure of this type mentioned above, the inside of the ball was considered to be at low pressure, in this case to retain the ions against their repulsive electrostatic force. The instability leading to collapse of the structure on any deformation from perfect sphericity was offset by an unspecified electromagnetic field which would realign the boundary. The violent end of the ball was again held to be an implosion.

Heterogeneous mixtures of greater complexity have been suggested for the composition of ball lightning, as in the theory that electrical discharges in storms generate positive- and negative-charged gases which, with water droplets and solid particles, form ball lightning.[546] The mixture was presumed stable at the high temperature of formation, and changes in the chemical composition on cooling were credited with the later behavior such as the final explosion. In further consideration of this theory the storage of energy in ball lightning by ionization was suggested.[32, 213] Up to 150 joules/cm^3 might be held in this way if all the air molecules were dissociated and ionized to produce monatomic ions, but the assumption was also made that the fireball is not at a high temperature, based on the numerous reports of such globes from which no warmth was noted. The form of the electrical charge present as molecular ions, ionic clusters, and charged dust particles was held to result in a very slow recombination of the opposite charges which would account for the long periods of luminosity of ball lightning. The light from the fireballs was ascribed to corona discharges between nonuniform charge distributions in the ball, molecular recombination, and combustion of occluded gases. The formation of ball lightning from plasma was excluded on grounds of the long luminosity. Presumably if a plasma were involved, the light would disappear in a few milliseconds as with ordinary lightning. The capture of free electrons of low energy by oxygen in the air was credited with the rapid decay. The absence of elections as a result of this capture was held to prevent any role of plasma (meaning the ion and electron fluid) in forming ball lightning, an advantage since free electrons were said to give more rapid recombination.[213] That plasmas undergo rapid recombination at atmospheric conditions is certainly correct, but experimental recombination coefficients are not in accord with the suggestion made in this theory that conversion of the electron into a negative ion reduces recombination. Indeed, for air gases it appears that ion–electron recombinations may be slightly slower than recombination between ions of opposite charge. Thus a large effect from the intervention of water or solid particles would be required to account for the long luminosity on this basis. When charge concentrations have already decreased to a low level and many neutral molecules are present, the greater diffusivity of electrons can give a larger number of recombinations. The storage of large quantities of energy in the ball by means of ionization requires high charge densities which are invariably associated with high temperatures. The constant appearance of almost all the globes in Rayle's survey,[420] in which no regular decrease in size or brightness or change in color was noted, was taken as evidence that dissipation of an initial store of energy which maintains the ball during its existence is not involved. The generation of a roughly spherical mass with a high charge density in which separation of charges is maintained for a long period against the rapid diffusion and recombination ordinarily expected has not been explained in these theories, which are thus equivalent in this respect to the older theories of the gaseous Leyden jar.

H. Vortex Structures

The problems of the formation of an isolated luminous mass in the sky and the moderate persistence of the resulting form combined with observations of ball lightning describing hollow globes, surface coronas, and rapid rotation led to theories depicting ball lightning as a vortex. Theories of this type were presented as early as 1859 when the formation of ball lightning by the collision of two oppositely directed strokes of ordinary lightning was proposed.[109, 337] This explanation has been suggested repeatedly by different investigators well into the current period, especially in connection with the reports of rapid rotation in some of the spheres (cf. Refs. 164, 216, 244, 260, 379). A sharp change in direction of the lightning channel causing a parallel but oppositely directed flow of the discharge for a short distance was also suggested as the source of opposing gas flows generating a rapidly rotating vortex.[322] The resulting structure was presented as a rotating layer of air at normal density surrounding an evacuated center, the centrifugal force of the rotating layer entirely balanced by atmospheric pressure. The generation of ball lightning in a normal lightning channel at the position where the descending leader carrying negative charge encounters the positive streamer rising from the ground was evidently considered a similar vortex-producing process involving oppositely directed streams of electrons and ions.[156, 183]

The escape of a hot jet of gas from the channel of ordinary lightning was also proposed as a method by which ball lightning is formed.[73, 74, 75] The magnetic sheath surrounding the channel during the high-current discharge of lightning would be weaker at the outside wall of a bend in the channel, resulting in the escape of a high-pressure electrically charged jet as if through a small hole. As the jet flows out through the magnetic field, it is rolled up into a ring or ball containing activated states of nitrogen and oxygen. These would decay slowly providing the desired long luminosity unless a disturbance such as the introduction of fresh oxygen occurs, causing an explosion by its sudden effect in increasing the decay of the activated states. The velocity of the jet of the dissociated and ionized gas at a discharge temperature of approximately $30,000°K$ in the lightning channel was estimated as 10^6 cm/sec, which is supersonic. Several liters of the gas could escape through 1-cm-square hole during the several-millisecond period of high current. This is sufficient to give a relatively large ball, 10 to 20 cm in diameter. Such a ball could contain 10^{11} ergs considering that the activated states may provide 2–12 eV. Lightning photographs with the Boys camera[74] show bright regions at the bends in a lightning channel remaining for 10^2–10^3 μsec compared to the usual decay of the light from the return strokes of approximately 10 μsec. An observation in which ball lightning seemingly originated in a sharp bend in a zigzag lightning flash[175] is also in accord with this theory.

Bent high-current spark discharges (100–2000 A) which were studied in gases at atmospheric pressure showed the formation of afterglows but of limited duration, inadequate to account for the long radiation time of ball lightning.[110, 111] Investigation of the spark channels showed that localized afterglows were not found primarily in the bends. While such long-period luminosity could account for ball lightning, the idea that bends in the lightning channel, possibly containing persistent ionization pockets, are brighter seemed to be ruled out. The discharge afterglows were attributed to excited energy levels of neutral molecules, considered more likely than ions and electrons.[111] Observation of brighter regions in lightning could result from sections of the lightning channel aligned with the line of sight. Such portions of the channel might be located at the bends which the observer can see in the path of the lightning running across his field of view. Later study of lightning channels, however, showed that bright regions exist at the bends for 10 to 100 times as long as in the straight portions.[74]

The generation of small vortexes constituting ball lightning in the atmosphere by whirlwinds, cyclones, or tornados was suggested[148, 149] following numerous observations of fireballs in connection with a tornado which appeared at night in France in 1890. Small spheres the size of billiard balls were seen moving rapidly with the whirlwind. One hit the ground and exploded. Bright red balls the size of a head slowly entered a barn, ignited the hay, and disappeared. Several saw luminous globes in their houses entering through chimneys or stove pipes. A large number of windows facing toward the storm were pierced with smooth-edged holes approximately 8 cm in diameter which appeared to have been melted in the glass.

According to this view ball lightning is the highly charged rotating sphere which separates from the lower tip of the rotating spout.[149] The electric charge was attributed to friction encountered at the high velocities by water droplets, hail, ice, and other solid particles in the storm. Whirlwinds known as dust devils, consisting of warm air rotating rapidly and entraining fine sand, occur especially in desert areas. A large example[161] approximately 8 m in diameter and 100–200 m high gave fields ranging from -450 to $+300$ V/m. The light from the ball according to this theory might thus be related to the blue-violet emission given off by charged water droplets in the experiments discussed previously.[331] The rapidly rotating gas was suggested as providing the wall lacking in the old theories which presented ball lightning as a Leyden jar. A later attempt to refute the existence of ball lightning, however, ascribed several observations of luminous globes to the light from whirlwinds in excluding the intervention of ball lightning.[197, 268]

The formation of ball lightning as a product of whirlwinds was related in the original theory to the supposed variation in the frequency of ball lightning with altitude.[149] Despite the great activity of electrical storms at high altitudes, ball lightning was said to be very rare there; this was ascribed to the absence of whirlwinds at great heights. This conclusion found agreement in observations from the South American Cordillère where several special forms of storm discharges were common but ball lightning was unknown[255] and, more recently, in Switzerland where no identifiable trace of this phenomenon was encountered by a meteorological investigator in 16 years of systematic storm observations.[43] Observations of ball lightning on mountain peaks, on the contrary, are actually reasonably numerous (cf. Refs. 117, 308, 413, 429, 490), and the opinion that ball lightning is more frequent in such locations is not uncommon.[461]

The appearance of luminous spheres in connection with the large rapidly spinning storms has evidently been long known,[145] and the balls may occur generally in the spout rather than only at the lower tip.[551] The possibility of a close relationship between the formation of ball lightning and tornados has been suggested.[53, 60, 551] The combined action of electrical and hydrodynamic forces in generating ball lightning as a highly ionized vortex was specifically adduced[85] in general terms in 1905. Frenkel described the spherical vortex as a mixture of small cloud droplets or dust with gas activated by the preliminary lightning flash in a structure resembling the Hill hydrodynamic vortex.[162, 277] The usual circulation of matter within the vortex, however, was replaced by alternating layers of separated electric charges of equal magnitude so that the vortex as a whole is neutral. The rapid motion of the charge layers is associated with magnetic fields which compress intervening shells of air. The structure is thus also a magnetohydrodynamic model, additional examples of which are discussed in plasma models of ball lightning. The energy of the ball is not large, perhaps 0.03 kW-h; the chemically active gases forming the ball react, releasing energy to the aerosol particles and thus heating them to glowing.

The picture of ball lightning as a cloud of ionized air constituting "vortical lightning" was used in an attempt to explain the effect of the Cougnard deionizer, a lightning protection device for high power lines.[139] The glowing spheres seen traveling on such a line were said to be formed by the attachment of the ionic cloud which is then propelled along the line by the wind. The protection afforded against storm damage and interference in performance by the device based on this theory was very effective although specific data on any involvement of ball lightning are lacking. An alternate process for the formation of the glowing masses traveling along power lines may be the corona or glow discharge caused by extreme overloading in the power surge following an ordinary lightning flash to the line, similar to the globules formed on passage of a current between two electrodes through a poorly conducting medium.[273] An increase in the external field around the line during a storm might also produce this effect with almost normal surges in a highly loaded line.

The generation of ball lightning entirely as the gas dynamic result of a cyclone with no connection to lightning[122] was proposed recently by an investigator who previously favored a nuclear reaction theory.[121] The observations of the globes bouncing from the earth were ascribed to gyroscopic rigidity and elasticity resulting from extremely high rotational velocity sufficient to give ionization of air through molecular collisions. The sphere was presumed stable as long as it maintained velocities above the ionization threshold. Experimental duplication of the vortex was suggested by means of ultracentrifugation or supersonic gas sources. Rotating masses of air produced by the firing of large guns were presented as low energy analogs of ball lightning in which the forward progress of the vortex is slow but rotation is rapid enough to tear leaves and branches from trees.[292] The far greater energy displayed in some cases of ball lightning involving the death of animals, the penetration of glass, the vaporization of metal, and the displacement of objects was attributed to the confinement of charged particles, evidently all of one sign, rotating in a ring. The magnetic field generated by rotation of charged particles in the ring, evidently in helical paths in its cross section as well as in circular paths around the ring, was suggested to assist in confinement.[292] The discharge of a large current through a fine wire ring was presented as a method of generating such a vortex. The characterization of ball lightning as an electron vortex was based on its formation over the nose of an airplane after the impact of linear lightning.[33] According to this theory ball lightning would form from the deformation of a charged gas layer on the surface of the craft. A stream of plasma directed against a parabolic surface resembling the nose of an airplane caused some deformation of the surface layer and possibly short-lived masses which may have detached themselves from the surface.

The vortex theories provide a very direct explanation of the numerous observations indicating rotation of ball lightning. The formation of luminous globes of this type can be readily ascribed to either the role of a preliminary linear lightning flash or the hydrodynamic action of a whirlwind. Such theories thus include a broad range in which ball lightning may be considered primarily an electrical phenomenon or entirely a hydrodynamic formation in which electric charges, if present, are additional results of the mechanical process. Some vortex theories may be classified equally well as electrical discharge theories and, especially in the most recent examples, as plasma theories. None of the studies of ball lightning as a vortex, however, has conclusively dispelled by detailed consideration the serious difficulties on which other theories have foundered, such as the continued luminosity of the balls for long periods while they travel inside structures. If the vortex is to be an isolated, self-contained sphere, its energy would be expended in viscous drag and turbulence in a very short time. A very few observations made at a distance and involving a faintly luminous region from which characteristic mechanical effects such as displacement of small objects occur may be ascribed to the intervention of small whirlwinds.

I. Ball Lightning as an Electrical Discharge

The explanation of ball lightning as a particular form of electrical discharge produced under favorable but rare conditions in nature has received the most extensive consideration of all the theories of ball lightning. In contrast to the chemical, ionic, and charged particle theories in which ball lightning is in a sense a secondary phenomenon composed of substances generated in a preliminary flash of lightning, this concept suggests that ball lightning is itself an electrical discharge or a specific region of one. Faraday early viewed this possibility with doubt on grounds that the behavior displayed by these luminous spheres was quite different from that of the electrical discharges known to him.[147] He held that the balls cannot be related to lightning or to any discharge of atmospheric electricity, because if electrical they should travel with high velocity and have a very short existence. Later investigators also expressed this opinion as a result of the dissimilarity of electrical discharges and ball lightning[480] and from reports that the glowing masses rolled on wet ground or failed to ignite combustible materials, indicating that they could not be an electrical discharge of any type.[164]

Soon after the publication of Faraday's opinion, however, the leading British lightning authority, Snow Harris, suggested that ball lightning is comparable to other forms of lightning in that it results from an electrical discharge, not one producing a spark but the brush or glow discharge of the type investigated by Faraday.[200] In the one and one-quarter centuries following this suggestion numerous experimental and theoretical investigations of these electrical discharges were made, often specifically for the purpose of elucidating the nature of ball lightning, and extended to studies of alternating current discharges. Later investigators restated Snow Harris' theory at intervals, and Faraday's opinion has also been repeated in different form and in greater detail.

In early experiments with an induction coil discharging in air to an insulating material such as glass or a varnished board wet with water, du Moncel produced reddish globes which he and others who repeated his experiments regarded as small models of ball lightning.[340] Du Moncel viewed ball lightning as the visible region produced in an electrical discharge from the cloud by the intervention of dry layers of air in the humid atmosphere. Such a discharge, he thought, occurred when there was insufficient energy for zigzag lightning, analogous to observations in experiments with conditions insufficient to give a spark over a wide gap. In dry air phosphorescence may be observed, but with humid air a moving luminous mass is formed. Du Moncel considered the spherical shape an aerodynamic result of the discharge through a resisting medium. He produced such a discharge in the middle of a candle flame. The hissing sound often reported with ball lightning was associated with the sound from a brush discharge. Du Moncel suggested the possibility that the electricity involved in generating the ball could be discharged in the form of ordinary lightning causing the ball to disappear.

In another theory a mechanism of formation was proposed which would indicate a close relationship between ball lightning and zigzag lightning. According to this theory a glow or brush discharge, the tip of which is most luminous and constitutes the ball lightning itself, is generated by a protrusion from a storm cloud.[385] The observed motion of the glowing sphere was held to be the propagation of this discharge rather than actual motion of any object. The transmission of electricity to the advancing fireball from the cloud is aided by the presence of water drops. The arrival of the ball lightning at the ground completes a conducting path along which a flash of ordinary lightning may instantly occur. In the terms of later descriptions of lightning processes, this theory depicted ball lightning as an especially luminous front on the advancing leader.

Experimental study of electrical discharges in gases at reduced pressure and in the air continued in attempts to duplicate the impressive natural displays seen during storms. Spherical forms were obtained in a low-pressure gas tube with an induction coil.[588] The formation of a long spark in air by electrical machines capable of generating greater potentials was usually compared to an ordinary lightning flash, but in an investigation of discharges produced under varying conditions by such a machine the glow produced on the positive electrode was favored over the brush discharge as a model of ball lightning.[539] A thin strip of wood was tied to the negative pole so that its two ends were both directed slightly toward the positive electrode, a 1-in-diameter brass ball. On application of the potential two bright spots appeared on the positive pole which remained stationary when this pole was rotated. When the negative pole with its strip of wood rotated, the position of the bright glows on the positive pole changed accordingly. This observation was related to the formation of ball lightning by the investigator who compared the tip of the wooden strip to a point in a negatively charged cloud. Motion of the cloud point during a flash would presumably cause the glowing spark on the earth to move in the manner of ball lightning. Observations of discharges from grounded conductors, such as water taps, which resembled the brush discharge and occurred following ordinary lightning were often reported as ball lightning and were a basis for the continuing view that ball lightning is this transition between the glow and spark discharges.[519, 548]

The studies of Planté beginning in 1875 with electrical discharges produced between insulated condenser plates by larger and larger lead storage batteries were directed primarily at the problem of ball lightning. Small luminous globes were formed initially with a 40-element battery and later with 1600 cells providing approximately 4000 V. At the lowest voltages investigated the small spheres were produced between platinum wire electrodes in salt water.[387] The glowing spheres were in continuous motion and emitted a crackling noise, often ending with an explosion. These experiments were also offered in explanation of the sound from cyclones and of the formation of fireballs at the tip of the spout. With the larger batteries a discharge tracing the form of a spherical surface was observed between an electrode wire above a water surface or a metal disk covered by a wet filter paper.[390] The discharge moved slowly across the flat surface following the motion of the wire electrode. Similar spherical discharges were produced when the battery was connected to condensers with mica or wet paper disks. These spheres were up to 1 cm in diameter and lasted for 1 or 2 min. The discharge occasionally bored holes in thin-layer mica or hardened rubber condenser plates.[390, 391]

The discharges produced in these direct current studies in many ways resembled those observed in the earlier work of du Moncel[340] and the later studies of von Lepel.[278] Planté suggested that studies with the voltaic currents could eventually show the electric charge of ball lightning which he anticipated would be positive.[387] His views on the nature of ball lightning changed greatly in publications over a period of ten years. Following an initial suggestion that the natural spheres are simply some electrically charged matter formed by storm discharges in moist air, he indicated that a condensation of the positively charged substance took place in a spherical envelope of rarefied air, metal vapor, and decomposed water vapor; but the experiments with wet condenser plates alone indicated that the metal vapor is not an essential component. Planté proposed that ball lightning is generated by a slow and partial discharge from storm clouds containing a great quantity of electricity when a portion of the cloud or a humid highly conducting column of air approaches the ground.[391] This column plays the role of the platinum electrode of the experimental studies, and the need for a highly electrified cloud is based on the more effective formation of globular discharges with long life when larger battery systems were used. The electric discharge formed under these conditions, as in the laboratory experiments, was considered a "sort of electric egg without the glass envelope," reminiscent of the Leyden jar structures discussed earlier.[391] The luminous mass was considered not dangerous in itself, as indicated by the great effect of the slightest air current on the laboratory globules and the behavior of ball lightning in its benign appearances. Its appearance would be a warning signal of great danger, however, for the position of the sphere indicates the most favorable path for discharge of the storm cloud's electricity, which is already partially discharging in the ball. The role in Planté's theory of the conducting channel from the cloud transmitting an electric current to the fireball anticipated the use of this mechanism in later more widely accepted theories, such as Toepler's, which differed greatly in other aspects.

Planté's experiments were considered a major contribution to the understanding of ball lightning[453, 565] although it was recognized that major problems remained, including the shape of the ball, its entrance into houses, and the identity of the natural equivalent of the laboratory wire electrode. In an observation of a whistling fireball following a lightning discharge the role of an iron-barred window and door in providing a shaped electrode for formation of the ball was suggested.[545] The further investigations of Righi and others with similar discharges[368, 432] produced large luminous masses displaying the low velocity and long life associated with ball lightning. The glowing form in Fig. 20, the only published photograph designated artificial ball lightning,[527] was generated in a repetition of Righi's discharge experiments. The larger discharge in the photograph exhibits a partial cloudy halo surrounding the central mass of the glow. Such a region has been reported in several observations of ball lightning in nature.

Continued studies of electrical discharges with the positive electrode a plate covered with an insulating material, as in du Moncel's work, produced similar red spheres.[386] With Crookes' experiments on a discharge flame in

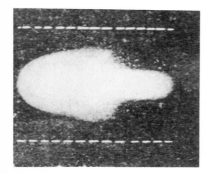 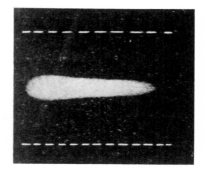

Fig. 20. Luminous masses formed in electrical discharge experiments. [J. Trowbridge, *Sci. Amer.* **96**, 489 (1907).]

air produced between electrodes in the secondary of an alternating current induction coil[113] as a basis, Hesehus studied[207, 208] the effect of a 10,000-V alternating current discharge with different electrode materials and shapes. Metal plates, water surfaces, water columns, and wet sponges were used as electrodes. Between a copper plate and a water surface 2 to 4 cm distant a very mobile discharge of varying shape was produced. Sometimes there was a conical flame; sometimes oval or ball shapes formed, changing with time. The color varied with the potential from red and yellow to blue, violet, and white. The fireball was never stationary but oscillated back and forth across the plate while emitting a crackling noise. Often it separated into several fragments which then reunited. No heat was felt from the discharge at a distance of a few inches. Fumes of nitrogen dioxide collected in a glass bell jar placed above the apparatus. The ball-shaped discharge was formed especially readily between a point electrode and a water surface separated by a short distance. According to Hesehus' report, decreasing the distance slowly after the ball was formed resulted in its disappearing quietly; and when the distance was increased, there was a sharp crack or explosion as the ball changed into the usual spark. Having reproduced a large number of the characteristics displayed by ball lightning in these experiments, Hesehus concluded that ball lightning is this particular form of an alternating current discharge of atmospheric electricity giving a nitrogen flame and producing oxides of nitrogen. Discharges resembling bead lightning were formed with the water columns. The discharges investigated in these studies were evidently not markedly different from those produced in earlier experiments and indicated that the alternating current generated luminous spheres closely resembling those reported by Planté with similar electrodes using direct current. The generation of ball lightning in nature by an alternating current was considered unlikely by de Jans.[230]

Luminous globes were formed at the negative pole when two fine metal points connected to an electrostatic machine were placed on the sensitive emulsion of a photographic plate.[226, 273] The detachment of such a ball from the very luminous negative pole left that point dark, and the small globe traveled slowly on the plate in a complicated path to the positive pole, sometimes pausing momentarily on its way. From 1 to 4 min elapsed as the sphere traversed the 5 to 10 cm distances between the poles. Sometimes the ball divided into two or three parts which were also luminous, and each then continued to the positive point. On arrival of the ball at the other electrode the luminosity disappeared, and the electrical source behaved as if the two poles were connected by a conductor. Development of the photographic plate made the path visible. The experiment also succeeded on a plate fogged by exposure to light which did not produce the conductivity in the sensitive emulsion until after passage of the luminous balls.

In a detailed series of experiments with direct current electrical discharges Toepler greatly extended the work of Planté in providing a basis for a theory of ball lightning.[514] Five types of discharge were distinguished with increasing current provided by the electrodes in air: the dark discharge; the glow discharge; the tufted or brush discharge; Toepler's brush arc, a striated discharge with dark layers separating the luminous regions; and Davy's Voltaic arc, the flame-like discharge. Toepler concluded that all the possible forms of lightning are brush discharges differing only in current[517] and that the striated brush, a transition between the tufted discharge and the Voltaic arc, most resembled ball lightning. The position of the glowing region between the electrodes once formed could be changed by varying the current supplied. In their motion between electrodes the discharges passed around plates placed in the direct line between the electrodes or through small holes in the obstructing plates.[515] Toepler studied the luminosity of the discharge as a function of the current in his experiments and concluded that as little as 0.01 A could give a very weak light the size of a child's head.[514]

Toepler suggested from these observations that the generation of ball lightning in nature takes place in the conducting channel remaining from a previous flash of ordinary lightning. If additional current enters the still-conducting channel from other parts of the storm cloud, the segmented discharge can occur, forming either bead lightning or ball lightning. The current required to give an easily visible discharge according to the experiments just cited would be a few amperes. Ball lightning is thus indicated to be of very low energy; only the initial lightning or a flash occurring later in the same path can account for any destruction. The color of the ball lightning would depend on the current in the channel as in the experiments in which a weak current gave a bluish glow while increasingly stronger currents gave dark red, brick red, orange red, and finally white.

The motion of the natural sphere is explained by two effects. The usual descent from the sky to the ground occurs in the channel as a result of a current change, and displacement of the discharge channel by the wind accounts for horizontal motion. If the surface of the earth in the role of the cathode[65] is the better conductor, the luminous globe is relatively quiet and stable until its disappearance, which may be accompanied by an explosion. If the cloud is the better conductor, the ball is very unstable and easily disturbed; it bounces and emits sparks. The luminous globe disappears if the electric current is depleted, for example, by the occurrence of an ordinary lightning flash from the same cloud source. Observations supporting this process have been reported.[452] The importance of the role of the lightning channel presented in Toepler's theory also received some support from the observation described earlier of a ball which formed directly under a cloud and moved slowly to earth in a channel which had been followed previously by several lightning discharges including one which decayed into bead lightning.[51] The elongation noted in several cases of ball lightning in which tufts appeared on the top or the bottom of the sphere giving heart-shaped or pear-shaped forms is in accord with the presence of the vertical conducting channel. As in Planté's theory, then, the glowing sphere is itself not dangerous, but it shows the position of a possibly dangerous path likely to be followed by a succeeding lightning discharge.[518]

Reference to ball lightning as some form of a brush discharge can be traced back at least to Snow Harris, and a similar suggestion was made by Oliver Lodge, who presented the possibility that underground metal conductors might be the source of such discharges in the presence of storm potentials.[288] From experiments with condensers in gases at reduced pressure the ionization of rarefied channels of air, which would be nonluminous, was proposed as a modification of Toepler's theory.[527] An increase of pressure in the channel would then produce the luminous region.

The initial effectiveness of Toepler's theory in explaining several characteristics of ball lightning was soon followed by recognition of observations which were beyond the capabilities of the theory. Some examples of ball lightning were reported in which there was no initial lightning immediately preceding the appearance of the globe or in which the sphere was formed at some distance from the ordinary flash (cf. Refs. 181, 229, 357, 360, 413). In such cases the presence of a nonluminous channel or invisible discharge such as that mentioned above would presumably be involved. The generation of only a single luminous region in the conducting channel is required in contrast to the experimental discharges which displayed several bright regions, as de Jans and Brand pointed out.[65, 230] The long existence of ball lightning would require that the channel provided by the initial lightning remain conducting for an equally long period. This would be a much longer time than ordinarily expected, but multiple lightning observations along a single path such as that described previously[51] indicates that it is possible.

Experimental studies show that very low continuing current maintains the so-called ball-of-fire discharge given by the mercury-pool arc against the normal microsecond decay which results when the current supplied is completely discontinued.[205] This type of discharge occurs at low pressure with low boiling point metal electrodes and is usually considered a self-sustained gas discharge. The potential drop between electrodes is approximately the ionization potential of the gas. In the ball-of-fire mode it appears that electron emission from the electrode is not thermionic but results either from tunneling, as from a high field at the cathode, or because of a decrease in the cathode work function resulting from vapor or plasma of high density near its surface. The electrons have only thermal energy in the plasma. At least 50 times as much current is carried by electrons in the arc as by ions. A similar discharge form is given by low voltage arcs with externally heated cathodes. The electrode potential in this case may be below the ionization potential of the gas.[234, 300] The discharge is maintained by the thermal energy provided by the cathode. The potential is a maximum in the ball where all the excitation and ionization occur. A typical ball 1 mm in diameter, a blue core surrounded by a pale rose shell, is formed[300] in argon at 6 mm Hg pressure with an electrode separation of approximately 1 cm. At higher pressures, such as atmospheric pressure, however, the formation of a ball in just this way would presumably not occur, as indicated by the approximately constant experimental value of the product of pressure and critical electrode distance for a given gas at pressures up to 13 mm Hg. Study of lightning flashes with particularly long-lasting light (0.04–0.27 sec) showed that continuing luminosity occurred when there was continuing low current maintaining the conductivity of the channel in discharges with multiple strokes.[253]

Formation of ball lightning as a discharge in the conducting channel of a lightning flash according to Toepler's theory should result in dimensions comparable with the diameter of the channel.[65, 230] Although many observations of ball lightning are in agreement with the typical diameters of 3–12 cm observed for lightning channels,[144] several are not. For example, in the same storm displaying a high frequency of lightning flashes in which ball lightning was observed following the channel of a previous flash, bright spheres three times the diameter of the lightning itself were formed at the lower end of flashes late in the storm.[51] Bead lightning moves very little either horizontally or vertically, and the beads maintain their relative positions. Toepler's theory would require the absence of wind or else uniform motion of the air along the entire chain of beads.[458] Ball lightning, on the other hand, may move extensively. One investigator concluded on this basis that Toepler's theory applies to bead lightning but not to ball lightning, which is a markedly different phenomenon.[354] The flight of ball lightning into houses would result from travel with the wind according to Toepler's theory. Motion against the wind reported in some observations is not explained. The rotation of the spheres and their rolling along the ground is not considered specifically. The generation of ball lightning in a clear sky[579] and in closed rooms were often cited as difficulties for this explanation. The observation of St. Elmo's fire indoors was noted by Toepler in support of the possibility that a brush discharge could also be formed.[514] Fifty years later ball lightning was again described[440] as an electric discharge from a point, similar to St. Elmo's fire but formed when an unusually strong field occurred. The reports[468] of St. Elmo's fire being transformed into a moving globe and leaving the fixed discharge point, which is absolutely essential for the initial corona-discharge form, provide an interesting basis for this theory. The objections to Toepler's theory led some investigators to the conclusion that the brush discharge is not applicable to the problem of ball lightning, and the studies of Rayleigh with active nitrogen produced in a discharge were suggested as an alternative.[480] The active nitrogen would provide a luminous mass with a light continuing after its departure from the electrical discharge. It was noted, however, that Lord Rayleigh did not accept this explanation of ball lightning.

Fifty years after Toepler's initial work consideration of the problem of ball lightning was directed again to the direct current discharge, in part because of theoretical difficulties of confinement in plasma models. Further experimental study was made of the discharge between a positive point and a plane or ring electrode.[483] With the exception of the usual corona phenomenon the only stable discharge obtained was a linear one. No experiments were made with wet or insulated electrodes such as those investigated by Planté, and no ball-shaped discharges were reported.

Further theoretical study of ball lightning considered a spherical, low current, dc glow discharge in the electric field following a linear lightning discharge.[151] The sphere is a region of high conductivity in which current and the lines of force of the electric field converge from wide regions above

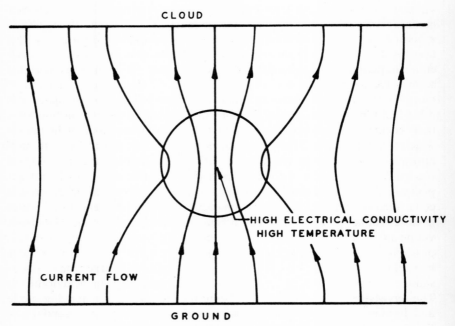

Fig. 21. Direct current discharge model of ball lightning

and below the glowing ball (Fig. 21). The conductivity is dependent on the square of the field. The short range dipole field of this model includes no force to keep the ball at a fixed height against buoyancy which would be present in a ball warmer than air temperature.

Luminous spheres formed in an rf discharge and considered possible experimental models of ball lightning[405] showed temperatures of 2000–2500°K. The effect of such temperatures was investigated with helium-filled soap bubbles which approximate the density of air at 2200°K. The soap bubbles themselves contributed little weight since they fell very slowly when filled with air. Helium-filled bubbles 20–35 cm in diameter rose in air, attaining a maximum velocity of approximately 120 cm/sec. The larger bubbles oscillated more into oblate spheres, and these flattened bubbles encountered greater drag which prevented them from reaching higher velocities. A ball of 20 cm diameter would require approximately 10,000 W to compensate for loss of hot gas at this rate.

To provide a structure in which the buoyancy of the ball could be overcome and the glowing discharge held at a stable height Powell and Finklestein modified the dipolar glow discharge theory to that of a positive-charged globe.[405] The fireball is said to originate as the residue of a stroke of ordinary lightning or through transformation of St. Elmo's fire. The theory deals with the processes by which the sphere continues for the long lifetimes reported after its appearance. The more rapid moving electrons leave the ball causing it to acquire a positive charge. The electric field near ground reverses after the preliminary lightning and thus exerts a force in the direction required to counteract buoyancy.

The rf discharge spheres generated in a resonant cavity remained visible for 0.5–1 sec after the power was shut off.[404] Luminosity continuing for such a long period after power is off has not been shown experimentally for similar dc discharges. This duration was taken as the life of ball lightning from the internal energy with which it is formed, and power from the reversed dc field of 2000 V/cm was proposed to sustain the discharge for longer periods.[405] The reversed field, which must continue for several seconds after the initial lightning stroke, provides additional ionization. A major process in the increase is found in Townsend multiplication from electron collisions in the field:

$$e^- + A \rightarrow A^+ + 2e^-$$
$$A = O_2, N_2, NO, \text{ or } O$$
$$A^+ = O_2^+, N_2^+, NO^+, \text{ or } O^+$$

Experimental studies have shown that in air an electron path of approximately 7 cm is required to give one additional electron in a field of 2000 V/cm by this process. In air with the equilibrium composition found at 3500°K less than 1 cm is sufficient to produce the additional electron when additional ionization processes also occur. A composition favoring electron multiplication would presumably be formed in air by the preliminary lightning stroke invoked in the theory. The temperature must remain above 2000°K for appreciable multiplication.

Electrons are channeled into the luminous ball from a wide region below it, as indicated by the upward current flow in Fig. 21. Positive ions leave the ball flowing downward in the field which maintains the position of the positively charged ball. The motion of such a charged sphere will be affected by the attraction of conductors. This explanation of the path of ball lightning is similar to the one proposed in a plasma theory in which escape of electrons as a corona current from the natural fireballs was considered to generate a positive charge.[128] An explosion may result on contact with a conductor from rapid heating by the sudden increase in electron current, increased breakdown of air may produce a lightning stroke, or the metastable molecules of the ball may suddenly release their energy.

A one-dimensional model of the current processes in the ball was studied with the following parameters:

Concentration of positive ions

$$\frac{\partial n_x^+}{\partial t} = \alpha_x \left| J_x^- \right| - \frac{\partial J_x^+}{\partial x}$$

Concentration of electrons

$$\frac{\partial n_x^-}{\partial t} = \alpha x \left| J_x^- \right| - \frac{\partial J_x^-}{\partial x}$$

Townsend multiplication

$$\alpha_x = A \exp(BE_x)$$

Poisson's equation of the electric field

$$\frac{\partial E_x}{\partial x} = (n^+ - n^-)/\varepsilon_0$$

Electron flux

$$J_x^- = -\mu_e n_x^- E_x$$

Positive ion flux

$$J_x^+ = \mu_i^+ n_x^+ E_x$$

The variables are n, the ion or electron concentration as indicated by the superscript $+$ or $-$; x, the position in the discharge; t, time; J, the ion or electron flux; E, the electric field; μ, mobility (~ 26 cm^2/V-sec for positive ions, 4 × 10^3 for electrons); ε_{02}, the permittivity of space; and A and B, empirical constants for the Townsend coefficient α_x, the number of electrons formed by an electron path of 1 cm in the field. Numerical solutions were obtained with the boundary conditions

$$J^-(x = 0) = 10^{11} \text{ cm}^{-2}\text{sec}^{-1}$$
$$J^+(x = R_0) = 0$$
$$n^+(t = 0) = n^-(t = 0) = 0$$

and the external field

$$E_0 \equiv [E(0) + E(R_0)]/2$$

which is also the field at the center of charge of the layer of thickness R_0. The field in the one-dimensional model extends to infinity because of the charge on the ball. Calculations were made for the discharge at one atmosphere and 3000°K in an external field of 1,750 V/cm with electron current of 10^{-8} A/cm^2 entering the layer. A time interval of approximately 10^{-4} sec, corresponding to one ion transit time of the discharge, sufficed to reach a steady-state solution. There is a rapid decrease in the electric field from approximately 3400 V/cm to 60 V/cm with an increase from 0 to 4 cm in thickness of the discharge layer as a result of the positive-ion concentration. With thickness greater than 4 cm the electron and positive-ion concentrations are almost equal, and further effect on the field is small. The currents from the ball have also become constant at this thickness, with slightly less than 10^{-7} A/cm^2 of positive-ion current flowing down toward the earth and 10^{-5} A/cm^2 of electrons emitted upward.

The equilibrium size of the sphere with respect to power losses shows stability against temperature changes, the radiative losses being proportional to the fourth power of the temperature and conductive losses to the first power, while temperature affects power input largely by a linear increase in the charge mobility. Radiation provides a stabilization process in this case. For variations in radius the significant parameters are the power input, radiative loss, convective loss, and mixing. The radius is stabilized by convective loss in much the same way that radiation stabilizes the temperature, and the radius and temperature are determined by given conditions, most important of which is the external field.

The light from the glowing spheres was described as radiation from electroluminescent air: long-lived metastable species produced by electron collisions in the post-lightning field which slowly transfer their stored energy to short-lived visible radiating species such as carbon dioxide. The molecular oxygen species $O_2(a^1\Delta_g)$ and $O_2(b^1\Sigma_g^+)$ were suggested as most likely long-lived substances with radiation lifetimes of 45 min and 8 sec, respectively. Both are stable at 2000°K. The first has been observed in quantities greater than 10% in microwave discharges, and the second is the source of strong emission in carbon monoxide–oxygen explosions at a pressure of several atmospheres. Estimates of collisional deactivation of these molecular states in air at 1 atm and 2000°K, however, indicated rapid disappearance in 1 sec and a very small fraction of a second for the respective oxygen species, a process which particularly excludes the metastable nitrogen species $N_2(A\,^3\Sigma_u{}^+)$ from consideration. The rf discharges gave unusually bright radiation at 2000–2500°K which would be emitted by air, if from species in thermal equilibrium, only at a higher temperature, possibly 4000°K. The out-of-equilibrium concentration of excited states according to this theory is produced by the collision processes in the post-lightning field. Radiation of approximately 50 W from the rf discharge appeared appreciably bright. Considerably greater power input would be required to maintain the excitation processes and the temperature, particularly if mixing with the exterior gases occurs. The visible light from the rf discharges in air consists almost entirely of radiation from the carbon dioxide continuum. Schumann–Runge oxygen bands, OH bands, and PtO bands are also recorded. Thus, organic impurities in the experimental system are probable sources of the light observed through short-lived species excited according to this theory by the metastable oxygen states. These oxygen species are unobserved in the rf discharges, presumably because their radiation times are longer than the period of the visible radiation from the discharges after power is off or because energy transfer to the short-lived species is favored.

The long-lasting light reported in the zigzag lightning which was converted to bead lightning[68] was ascribed to the presence of the metastable excited species invoked in this theory.[405] In study of lightning flashes with long luminosity, however, current continued in the channel between multiple strokes, indicating that direct excitation of the visible short-lived species may account for the extended luminosity.[253] Similarly, the direct excitation of such species may take place in the suggested long lasting electric field of reversed polarity following ordinary lightning without the agency of metastables. Glowing spheres of long life after discontinuation of the power have not been reported from dc discharges.

The existence of ball lightning in airplanes for longer than the decay period associated with the experimental rf discharges was attributed to the excitation of rf resonances in the aircraft fuselage by an external dc discharge.

A similar model of ball lightning was investigated with both thermal conductivity and electrical conductivity nonlinear and functions of the temperature rather than the field or the current[532]. Such a globe would be in a region of hot air as shown in Fig. 21. In the thermal model the temperature is maintained by external electrical power, and only thermal conduction is considered as a mode of power loss. The initial condition suggested for formation of this discharge is a volume of heated air, possibly the conducting and cooled channel remaining from a flash of lightning as in Toepler's theory. A ball formed in this way with a 20 cm diameter and a central temperature of 5000°K is described as having an energy of 2000 joules and radiating 100 W of visible light to give the approximate luminosity of a 1000-W light bulb. The time required in cooling a ball of this size containing air initially at several thousand degrees with no further supply of energy was calculated as several seconds.[535] Heat loss was permitted only by thermal conduction, with the usual energy and momentum balances and mass conservation in the ball. The pressure in the globe was assumed to be atmospheric, and no pressure changes were considered. The luminosity of the globe would decrease with cooling from the 5000°K incandescence, an order of magnitude or more with each thousand-degree decrease in temperature, dropping to 3000° in approximately 1.5 sec. The investigator pointed out, however, that in many observations the light from the fireballs appears constant rather than decreasing as this model would have it. The hot sphere would also rise rapidly in the absence of the large force required to hold it in position against the great buoyancy of such a ball in air. No confinement of the high-temperature gas to the region of the ball is evident in any case, since the absence of even a retarding pressure increase to the surrounding air would permit the high-velocity gas molecules to diffuse out rapidly.

These later theories of ball lightning present initial attempts at analytical descriptions of the electrical discharges which have long been suggested as models of ball lightning. The difficulties encountered by the discharge theories in accounting for observed characteristics have not yet been mitigated by these studies. The total energy estimated for such spheres is 10^{-3} of that derived for a high-energy ball in the incident of a water tub heated to boiling, yet the temperatures cited indicate they are not the entirely cool balls from which no warmth is sensed.[184, 345]

The classification of ball lightning as some form of an electrical discharge, which provides a basis for relating it to other storm discharges, has received the most extensive consideration of all the theories of ball lightning. It possesses the signal advantage of utilizing an ample source of energy which all recognize to be present in storms. Some of the most striking characteristics of ball lightning, its motion, its appearance in enclosed spaces, and its failure to be affected by conductors in several cases, remain unresolved and present particular difficulty for this theory.

J. Luminous Spheres from Vaporized Substances

In his early discussion of lightning phenomena containing the suggestion that ball lightning may be a brush or glow discharge, Snow Harris also described the formation of red-hot globules of metal by electric discharges in wires.[200] Iron, copper, and lead wires produced mobile spheres which bounced off a cool surface repeatedly in Van Marum's experiments[538] published in 1800. Direct suggestions that ball lightning is not an electrical discharge but a luminous molten substance[524] or, specifically, incandescent metal[355] were made much later. Planté compared bead lightning formed from an ordinary lightning flash to a segmented incandescent chain produced by an electrical discharge in metal wire.[389]

Several observations of ball lightning indicate the possibility that an initial linear flash of lightning may produce a glowing sphere by vaporization on striking a solid material. The appearance of a ball of fire rolling away from a tree which has been struck by lightning is typical.[469] This may in some cases be a corona or glow discharge from the advancing front of the current flow, as in the globules traveling between electrodes on a photographic plate,[273] or it may be a high-temperature incandescent mass resulting from intense heating by the lightning current. Resin found in a fulgurite from ball lightning following a flash to a tree[328] is in accord with such a process, although the role of ball lightning in the incident is not clearly established by the report. A tar-like residue was reported in another incident of a burning ball on the ground which exploded after half a minute.[564] The fireball occurred during a violent lightning storm. The residue smelled like sulfur and was hot enough to burn the witness's hand 10 min later. His fingers turned yellow from touching it.

Glowing balls have also been observed following lightning flashes to metal structures, and in some occurrences vaporization of a length of telephone or antenna wire was evidently closely connected with formation of the fireball. A bright mass almost 1 m in diameter was seen rolling along the street accompanied by others of smaller size after a very heavy discharge of lightning to a telephone pole.[398] Two wire ends were found later dangling from adjacent poles with a considerable length missing. In another case a very strong flash of lightning destroyed a copper wire antenna soon after it had been installed for studies of storm electricity.[528] Witnesses viewing the storm saw a large ball of fire evidently formed by the vaporization of a 65-m length of the 2-mm wire.

An accidental discharge to a copper wire in a student laboratory produced a glowing ball similar to those reported by Van Marum which seemed to roll slowly across the table until it disappeared.[237] Along the path of the ball there was a line of scorched spots which ended at a crack in the table 1 or 2 mm wide. In a drawer directly under the crack was found a small copper sphere approximately 1 mm in diameter. The luminous globe was apparently copper heated to incandescence. A witness reported it was yellow-white rather than the green he would have associated with vaporized copper. The generation of evidently similar luminous masses by a short circuit in high-power electrical systems was described by Brand.[65] In one case a 5-cm-diameter ball was produced which traveled with the wind approximately 50 m. Green incandescent globes 10–15 cm in diameter were generated by large discharges as high as 4×10^7 W through copper and silver electrodes.[477, 479] The lifetime of the fireballs was reported to be on the order of a second. The initial energy of the ball was estimated at between $0.02–0.4 \times 10^6$ joules from assumption of a 5–10% conversion efficiency of the power in a discharge for 0.01–0.1 sec. Ionization of all the gas molecules in a 10-cm-diameter sphere at standard temperature and pressure, assuming 15 eV is required for production of each ion–electron pair, would store an amount of energy slightly above the lower energy estimated from the power conversion. To prevent the rapid recombination which would be expected with such a fluid an unspecified potential energy provided by the configuration of the ball was suggested. The charge density involved in this estimate of the ionization would be approximately 10^{19} ions and electrons per cubic centimeter, close to the 4.3×10^{18} electrons/cm^3 estimated in a lightning stroke.[531] Extremely rapid disappearance of the ionized ball would be expected then, both from the short life of ordinary lightning and from the rapid recombination observed with high-concentration ion and electron plasma fluids.

The separation of free-floating spheres from St. Elmo's fire and from the luminous balls traveling along power lines has been reported. This transformation would be ascribed to the conversion of the initial glow discharge into a brush according to Toepler's theory,[180] but the vaporization of a portion of the conductor, which is often a metal, as in the luminous masses just described, is likely with a long lasting or intensified discharge.

De Jans held that the characterization of ball lightning as an incandescent ball of metal vapor was incorrect and the observation of such a globe formed by lightning vaporizing a telephone wire was a special case not connected with the natural phenomenon.[230] The glowing balls obtained experimentally by intense electrical discharges through metal, on the other hand, exhibit the free motion and long life reported in ball lightning observations which present such difficulty for other theories.

The metal-gas or liquid spheres themselves involve further questions on the composition of the material and the physical structure of the ball which is formed. Preliminary consideration of high-temperature globes of carbon vapor indicated that only condensed phases could give density sufficient to approach that of air at normal conditions to give a nonrising ball.[293] The predicted radiation associated with either vapor or suspended condensed particles in such spheres was insufficient in intensity or duration to compare with reported observations. No satisfactory mechanism has yet been given for retarding recombination of ionic species. Ionization, if initially present in these spheres, may not be expected to persist for a time comparable to the life of the glowing mass. Estimates of the cooling of hot spheres by thermal conductivity alone indicated that a few seconds would elapse in cooling of a sphere 20 cm in diameter.[535] The exponential dependence of cooling time on the sphere size according to this theory, however, would give a lifetime of only a few hundredths of a second for the small copper globe produced accidentally as described above. Thus, thermal conductivity alone would also presumably dissipate initial high temperatures in a shorter time than observed in this case. The examples of the laboratory vaporized metal globes which persist for long periods indicate that the energy loss processes mentioned do not determine the life of these experimental bodies but that additional parameters must be considered.[482] If a low energy content, which is indicated as more common in observations of ball lightning,[420] is accepted, the light radiated may be produced by relatively low-energy, long-lived electronic excitation levels of the gases or the active nitrogen of Rayleigh. Transfer of the excitation energy from such species to the metal atoms may be involved. The persistence of the geometric form of the vaporized spheres indicates equally that a stable structure, for example, a smoke ring or a vortex,[332] is formed. Conclusive explanations have as yet not been offered for either the prolonged luminosity or the stability of the form of the vapor spheres.

K. Plasma Theories and Experiments Applicable to the Problem of Plasmoids

The unusual properties exhibited by ball lightning and the failure of repeated attempts to relate these remarkable bodies to other, better understood phenomena gradually led to the view that ball lightning is composed of an unusual substance to which distinctive terms were applied by different investigators. Ponderable matter, from the significant density shown by the incandescent fluid in its rapid descent from the sky,[524] globular vim,[130] and fulminating matter[310] are typical designations which were given to the presumably high-energy, electrically charged, luminous material. The continued appearance in ball lightning theories of globular structures containing high concentrations of ions and electrons was gradually related to the information being obtained on the production and maintenance of high-temperature fluids with high charge densities in nuclear physics. The material of such fluids is known as "plasma," and a geometric structure of limited extent containing this substance is a plasmoid. The experimental observations and theoretical descriptions of plasma physics are of great value to the problems of ball lightning, particularly in connection with the most recent theories in which ball lightning is generally regarded as a plasmoid.

The processes required for generation of a plasma, the properties of the plasma fluid, and the conditions of formation of a plasmoid are essential information to be supplied from the large store of plasma physics. In some of the questions involved a conclusive reply is not yet at hand. Ultimately the characteristics displayed by the plasma models must be compared with those so often reported in ball lightning. The topics to which the following general discussion is confined have been selected with this ultimate comparison in view. Plasma theory and experiments dealing explicitly with ball lightning are considered in a later section.

High concentrations of electrons and positive ions are suggested in some theories of ball lightning to explain the high energy content indicated by the long period of luminosity and the final explosion from a rather small mass. The plasma consists of the fluid of numerous electrically charged particles. Strong electrical forces prevent any substantial deviation from charge neutrality which might occur, for example, by greater concentration of a given charge in a region. Particles of the opposite charge rapidly flow in to neutralize the region. The greatest distance to which the field from a charge may be detected in a spherical volume of the plasma is approximately[285]

$$d = \left(\frac{2kT}{ne^2(n_e + n_i)}\right)^{1/2}$$

in which k is the Boltzmann constant, T the kinetic temperature, e the electrostatic unit of charge, and n_e and n_i are numbers of electrons and ions, respectively, per cubic centimeter, for ions only of unit charge. This is the spherical Debye shielding distance obtained from the energy of the plasma particles, in this case the thermal kinetic energy equivalent to the electric field. The ordinary random motion of the charged plasma particles may thus cause a separation of charge to this distance. Nonuniformities may be detected within this length; outside of it a uniform fluid description is generally applicable. An ionized gas in which the shielding distance is small compared to the total region involved, for example, the diameter of the plasmoid, is considered a plasma. In the case of a sphere such as ball lightning this requires a diameter of at least ten times the shielding distance if the material of the ball is correctly described as a plasma. The estimates of ionization in ball lightning and the pressures and temperatures suggested indicate that the plasma description is applicable and may be of interest even when the gases in the ball are only partially ionized.

The formation of a bounded structure containing the ions and electrons requires confinement within the plasmoid. A few general methods are known, primarily from thermonuclear research, by which such particles may be retained at moderate density in certain favorable structures. A strong magnetic field can be used, or a combination of magnetic and electric fields. A charged particle travels freely along a magnetic field, but if it moves across the field cutting the field lines it is deflected. The radius R of the resulting curved path of a particle with mass m and charge q crossing the constant magnetic field B with a component v of its constant velocity is

$$R = \frac{mv}{qB}$$

or in terms of the thermal energy

$$R = \frac{(2mkT)^{1/2}}{qB}$$

The electron, being of much smaller mass than the positive ion, is more readily confined. Comparing, for example, an electron and an ion with only the thermal energy from lightning with a temperature of 10,000°K (slightly less than 1eV), the radius of the circle described by the electron in the magnetic field of the earth is approximately 10 cm, and the radius of the molecular nitrogen ion N_2^+ is more than 22 m. The motion of the two particles may not be considered entirely independent in this way if they are in a plasma, for any charge separation greater than the Debye distance would cause electrostatic potentials greater than possible from the thermal energy. With a sufficiently strong magnetic field the paths of both particles can be confined within the dimensions of the plasmoid. An electric field may be added to assist in retaining the heavy ions when the magnetic field is weak. The fields may be either internal ones resulting from currents and charges in the plasma, or they may be external. Some combination of these is found in most plasma studies for thermonuclear purposes. The particles must follow a closed orbit, and drift of the orbit must be limited. Several variations of the magnetic bottle and the torus, or ring current, have been most extensively investigated.[276] Very strong external fields combined with carefully selected internal plasma flow have been found necessary to confine energetic plasmas for even a few microseconds.

Electromagnetic waves can also provide confinement. The motion of a given charged particle may be considered as in tune with the alternating electric field of the wave. After the particle travels a short distance, the field changes, returning the particle to its original position if the frequency and strength of the field are appropriate for the charge and mass of the particle. Confinement by an electric field alone in this way may be described[373] by the Mathieu equations in the form

$$a_x + \frac{2q}{m}(U + V \cos \omega t)\frac{x}{R^2} = 0$$
$$a_y - \frac{2q}{m}(U + V \cos \omega t)\frac{y}{R^2} = 0$$
$$a_z = 0$$

for cylindrical confinement of a beam by a quadrupole field with no change in the velocity of the particle along the cylinder. Such a field may be given by four cylindrical rods to which the harmonic potential is applied placed in a square around the focusing region containing the plasma. Adjacent electrodes bear a constant electric potential of the same magnitude but opposite sign, $+U$ and $-U$, giving opposite electrodes the same potential. The alternating field $V \cos \omega t$ now provides the acceleration components a_x, a_y to the particle with a charge-to-mass ratio q/m at position x,y in the plasmoid of radius R. The appropriate frequency and magnitude of the field may thus be selected for confinement of any charged particle.

Iron and aluminum solid particles of approximately $20\,\mu$ diameter with a charge of 0.01 coulomb/kg have been contained in this way.[586, 587] Radio frequency confinement was studied theoretically with a nonneutral plasma containing excess electrons, considering the additional second-order force from motion of the electron in the varying electric field. The coulomb force from the excess charge and a pressure gradient appeared necessary to obtain confinement.[55, 56] Further approximate investigation for an isothermal plasma sphere indicated that there would be appreciable confinement of a neutral plasma containing low-energy electrons when the plasma frequency is the same as the electromagnetic frequency.[116] This method was extended to include the effect of collisions and the electromagnetic force from the electron moving with the magnetic component of the wave.[167, 191] The electromagnetic force appeared less significant, but the Mathieu force was still appreciable for the special condition of plasma resonance with the applied wave. A dense plasma affects the fields from the electromagnetic wave strongly; the usual Mathieu solutions are not applicable under such conditions.

The use of the high-frequency field in combination with a constant magnetic field, similar to its use with the constant electric field mentioned previously, was proposed as more effective in confinement.[64, 540] The alternating field was utilized as an auxiliary force in focusing a cylindrical plasma. With dense plasma there is no penetration by the fields into the plasmoid to give strong focusing according to the Mathieu functions. The energy of the radiation field in the thin boundary layer gives, rather, a confining pressure balancing the plasma forces of the form

$$nkT = \frac{E^2 + B^2}{8\pi}$$

in which E and B are the electric and magnetic field components of the radiation. Experimental results showed confinement of a plasma with a density of 10^{13} particles/cm^2 containing 5-eV electrons ($T\sim5.8 \times 10^4\,°K$) by a 60-gauss magnetic component of a wave.[540] The field provides approximately 70 dynes/cm^2, which is in accord with the particle force.

In studies of a relatively low-density hydrogen plasma with on the order of 10^{10} electrons/cm^3, cylindrical confinement was obtained from a microwave electric field and a constant magnetic cross-field.[141] A weak microwave field supplying 0.2–1 W was sufficient when the plasma was at the appropriate resonance for this geometric structure

$$\omega_p{}^2 = 2\omega(\omega - \omega_b)$$

where ω_p is the plasma frequency, ω the frequency of the applied field, and ω_b is the cyclotron frequency of the electrons. The volume of the plasmoid was proportional to the microwave power.

Cylindrical structures formed by rapid rotation of charged particles in a uniform magnetic field directed along the axis of rotation were studied theoretically.[249, 547] In this case the confining force balancing the space charge and the thermal energies is obtained only from motion of the charged particles through the external field. The particle distribution function for the plasmoid structure was obtained in which the diameter is related to the temperature and plasma frequencies. Uniform rotation as in the motion of a solid body and nonuniform rotation as in a galaxy were considered. In rotation at the Larmor frequency $qB/2m$ a resonance occurs[249] when the ratio of the square of the ion plasma and the rotation frequencies $\omega_p{}^2/\omega^2 = 2$, corresponding to a resonance observed in the experimental work cited.[141]

The ring current and toroidal plasma has been studied extensively as a promising structure,[276] and the production of ball lightning in this form is often suggested. Small plasmoid rings set in free motion are generated by plasma guns with short high-current discharges in parallel wire electrodes [58, 59] or induction coils shaped to eject the plasma formed.[283, 569, 570] The effect of motion through magnetic fields and interaction of several of these masses directed at each other have been studied. A magnetic field is trapped in the plasmoid on formation, and there is a continuing toroidal current.[283] The maximum period of luminosity[59] until disappearance of the ring, either by complete recombination or diffusion of its substance, is less than a millisecond, even at a pressure of 10^{-6} atm. A very small increase in the pressure slows the plasmoid to a stop from initial velocities of 10^5–10^7 cm/sec.

The confinement in a torus of a high-density plasma of value for thermonuclear reactions is beset with difficulties.[185, 276] With a toroidal magnetic field applied to the plasma chamber, the field lines proceeding in the same way as the ring current in the plasmoids just discussed, retention of the charged particles against cross-field drifts requires somewhat complex field configurations. A rotational correction of the magnetic field, achieved either by twisting the torus into a figure eight or by adding to the ordinary torus helical windings in which alternate wires carry the current in opposite directions, is effective for limited plasma pressures or current. On the other hand, a stable equilibrium in an external toroidal magnetic field is possible for a low-density electron current for, in contrast to a neutral plasma, the magnetic field gradient and centrifugal drifts are not significant.[120] For example, the magnetic gradient can give only relatively weak electric fields in the absence of opposite charges.

Spherical plasmoids of relatively small dimensions, such as those reported for ball lightning or involved in experimental work, have presented greater difficulty for theoretical analysis, and no general solution for this structure equivalent to the detailed consideration of the cylinder and torus has been presented. The isothermal gas sphere in gravitational equilibrium has long been of interest in astrophysics. The equilibrium radial distribution of the mass is given by[98]

$$\frac{1}{R^2} \frac{d}{dR}\left(\frac{R^2}{\rho} \frac{dP}{dR}\right) = -4\pi g\rho$$

when the total pressure P is a linear function of the density ρ at a given temperature; g is the gravitational constant. If an exponential function $ae^{-\psi}$ is selected for the density, this may be transformed to

$$\frac{1}{\xi^2} \frac{d}{d\xi}\left(\xi^2 \frac{d\psi}{d\xi}\right) = e^{-\psi}$$

in which ξ is proportional to the radius given in the previous equation.

An equilibrium gravitational sphere of a conducting fluid has been shown theoretically for a magnetic field which lies entirely inside the fluid. There is no magnetic field either on the surface or outside of the sphere.[408] The field lines are like a helix on the surface of a torus; both poloidal and toroidal components are present.

A plasma structure with a spherical boundary was suggested on the basis of an analogy between magnetohydrodynamic and hydrodynamic vortex parameters.[474] The magnetic field **B** of the plasmoid was compared to the velocity field **v** of a stationary hydrodynamic vortex, and the current density **j** to the vorticity Ω.

Hydrodynamic Vortex
    div **v** = 0
    curl **v** = Ω
    $\Omega \times \mathbf{v} = -\nabla(p/\rho + v^2/2)$

Plasma Vortex
    div **B** = 0
    curl **B** = **j**
    $\mathbf{j} \times \mathbf{H} = \nabla p$

Using a specific hydrodynamic structure such as the Hill vortex[332] the

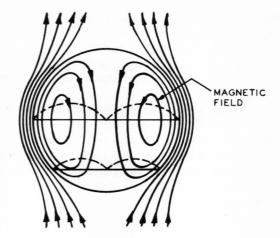

Fig. 22. Vortex plasmoid.

internal configuration of an equilibrium spherical plasmoid may be obtained in terms of the pressure, current, and magnetic field components. This sphere is analogous to a ring current in an external magnetic field directed along the central axis of the ring (Fig. 22).

The role of gravitational force in the astrophysical spheres is essential in giving the equilibrium structure. In the absence of this internal force the difficult problem arises of providing confinement fields with equivalent symmetry. Complex combinations of the external pressure, magnetic fields, and helical and toroidal currents become necessary.

Aside from the establishment of an equilibrium configuration in which a net steady-state balance of forces is provided, instability of the plasma fluid in transient fluctuations from the equilibrium is a major difficulty. The desired structure may be entirely obliterated by oscillations of the plasma induced by the external or internal fields. Thus in confinement of a plasma cylinder by high-frequency electromagnetic waves, a steady magnetic field was used as the major source of confinement, and the alternating field only as an auxiliary. The instability generated by the electromagnetic field was minimized by damping provided by the constant magnetic field and by minimizing the region of contact between the wave and the plasma.[540]

The mobility of the plasma fluid combined with the electric and magnetic interactions of its electrons and ions can give rise to complex motion.[276] A few of the simple instabilities encountered illustrate the sources of this motion. In a plasma cylinder surrounded by a magnetic field parallel to its axis the plasma can move so as to extend outside the original circular cross section into the region of the surrounding field while the magnetic field occupies the

Theories and Experiments on Ball Lightning

area left by the plasma (Fig. 23a). This exchange of plasma and field positions (known as interchange, ripple, or flute instability) becomes significant when the field lines are not deformed and the plasma energy decreases or does not change with the oscillation. A plasma cylinder surrounded only by its self-magnetic field, the field arising from the current flowing down the tube, displays a sinuous motion, the kink instability.[262] When a short section of the cylinder is moved giving a curve, the magnetic pressure on the inside of the curve, the concave side, increases as the field lines are crowded together (Fig. 23b). On the convex side of the curve the magnetic pressure is reduced, resulting in a net force increasing the curvature and favoring the instability. The cylinder may be stabilized against this type of deformation by surrounding it with metal walls which trap the magnetic field and compress it. The sideway motion of the plasma toward the wall thus soon encounters a magnetic pressure increased by trapping to equal the force exerted by the field increased by curvature on the opposite side of the plasma cylinder. In the same cylindrical plasma column another deformation produces a narrow neck by the collapse of the outer wall all around the tube (Fig. 23c).

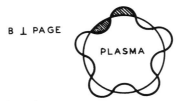

(a) Interchange, Ripple, or Flute Instability

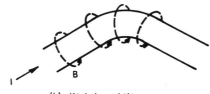

(b) Kink Instability

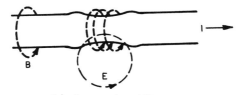

(c) Sausage Instability

Fig. 23. Simple plasma instabilities.

The curvature in the neck produces greater magnetic pressure in the thinnest region by crowding of the field lines, giving greater narrowing. The rapid collapse of the magnetic field at the neck in turn induces a large electric field across the depression which can exceed the externally applied field. This field occasionally produces larger energy particles than expected in plasma experiments sufficient for thermonuclear reaction for brief periods. If a magnetic field is applied along the axis of the cylinder when the plasma is generated, the field is trapped, and an internal magnetic pressure resides in the cylinder. This field provides a stabilizing force against the sausage deformation which is greatest at the narrow neck. The instabilities described, among others, are evidently applicable to the lightning discharge, and a role has been suggested for such processes in formation of bead or ball lightning from the linear discharge.

Of the typical oscillations and deformations of cylindrical plasma wriggling is especially encountered in a ring.[474] Although maintenance of a plasma ring with an internal pressure greater than that outside was deduced from the possibility of using a magnetic field strong enough to confine any internal plasma energy, stability against the flute oscillations is obtained only for lower internal pressures.[266] Toroidal plasma experiments in a strong external magnetic field show expected stabilization against these oscillations when the radius of the torus and the discharge current are sufficiently small, even while the plasma ring is displaced in the experimental chamber by the usual cross-field drift.[132] The flute instability proves to be readily avoided, however, even in unfavorable fields such as a concave magnetic field, for plasma with a relatively low energy compared to the external field.[275] Other instabilities such as oscillations from charge flow along the magnetic field present greater difficulty.[276, 559] Specific conditions of pressure, magnetic field, and radial curvature of the toroidal envelope can also stabilize the instability from the field-parallel component of the current[559]; but the more basic question of the lack of existence of a true toroidal equilibrium under realizable conditions remains.[185] Numerous difficulties in the plausible structure of a ring current appear on comprehensive consideration of the plasma properties involved in confinement.

Theoretical investigation of a conducting sphere assumed to be an equilibrium structure in an external electromagnetic field led to the conclusion that stability against small deformations is possible.[589] The external electric field causes instability in general, and the magnetic field at the surface must be larger by a factor dependent on the nature of the deformation. In a uniform external magnetic field, concentration of the field in wells and spreading at protrusions in the surface of the sphere increases the deformation. Deformations of twisting, however, increase the magnetic pressure resulting in general stabilization. The oscillation of the complete field, suggested as the product of three component waves, prevents correlation of the instability with the field. The field rotation stabilizes the sphere against small deformations. The model used was incomplete with respect to specific plasma fluid parameters essential in determining the desired structural characteristics, as indicated in consideration of the toroidal problem. The oscillations in a charge-neutral electron sphere have been studied with appropriate plasma velocity, pressure, density, and electric and magnetic field relations to obtain electromagnetic and acoustic modes.[590] The ion waves given by spherical plasmoids generated by 15-MHz frequency in a glass sphere were investigated. The resulting frequencies observed with the gases oxygen, nitrogen, hydrogen, neon, and mercury ranged from 82–260 kHz and were approximately dependent on $(3kT_e/m_i)^{1/2}$, involving the electron temperature T_e and the ion mass m_i.[172]

The formation of luminous plasmoids as electric discharges produced by alternating current has long been of interest.[582] Such high-temperature masses are often studied in gases at reduced pressure in glass chambers in the field of a coil or between electrode plates with no direct attachment to the electrical elements and are thus known as electrodeless discharges.[201] The applied field for those masses of low charge density can directly control electron motion to produce ionizing collisions of gas molecules and generate the plasma region. A wide range of frequencies (for example, from 50 to 10^8 Hz) is effective.[27] The qualitative properties of these discharges obtained with power from 10^{-2}–10^3 W/cm^3 and up to atmospheric pressure were reported. Cold discharges with microampere currents and arc-like discharges with hundreds of amperes and kilowatts were produced. The discharge colors in atmospheric gases with megahertz frequencies vary with pressure from pink-red at a few millimeters, yellow-white at above 10 mm Hg, to fiery orange at several centimeters pressure. With 62-MHz power the color changed from pink-red up to 5 mm Hg, to purple or red-brown, green, and finally yellow-white above 400 mm Hg. These discharges were interpreted in terms of different types of current flow presumed in the luminous region rather than the composition, energy, and distribution of charged particles of interest in later plasma studies.

Megahertz frequency discharges have been used as hot torches.[106, 421] An induction coil surrounding the gas flow tube produces a plasma flame stable at atmospheric pressure once ignited. Temperatures up to 19,000°K and 3.1 kW of power were available in the torch plasma.

A luminous, pale blue disc rotating rapidly was generated at low pressure in the cylindrical apparatus shown in Fig. 24. The discharge formed in an electric field of 100 V/cm at 0.5 mm Hg pressure, a magnetic field of 6000 gauss along the cylinder axis giving the rotation. The rate of rotation was 17,000 rps. Gas velocities of 1.3–2 × 10^5 cm/sec were measured by injecting

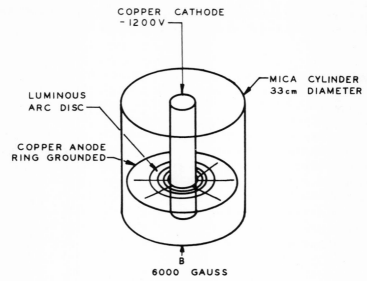

Fig. 24. Supersonic rotating arc.

a colored gas and by determining the force on a tungsten vane placed in the flow. The discharge gave low temperatures, less than 600°C, although appreciable power, 0.5 kW/cm, was applied.[136]

The storage of energy in a similar rotating plasma, the "hydromagnetic capacitor," has been reported.[10] Argon at 0.4 mm Hg pressure in a short cylindrical chamber with conducting walls and central pole was subjected to a rapid 6-kV discharge. In an 18,000-gauss magnetic field along the axis of the cylinder the plasma was set in rapid rotation with currents up to 200 kA flowing.

More complex fields for confinement of plasma such as octupole magnetic fields formed by four large hoops in a toroidal chamber with a square cross section have been effective in trapping a fluid of 100-eV ions and 10-eV electrons.[133] The density of the plasma gradually diminished, presumably through loss on the ring supports, with no evident instability. Magnetic levitation of the hoops can avoid the loss of plasma.[437]

The combination of detailed current flows and corresponding self-fields to produce self-confinement of a plasmoid has been occasionally suggested, especially for ring currents. Consideration of the total energy content of plasma, however, has repeatedly led to the conclusion that a nongravitational plasmoid cannot exist in any closed equilibrium structure.[474] This is a result of the conditions necessary for stable equilibrium of a closed plasma system, in which the gravitational, electrical, magnetic, and internal fluid energies must compensate each other as stated by the virial theorem

$$G + E + M + U = 0$$

If gravity is absent ($G = 0$), no equilibrium is possible in a closed system according to this limitation. The possibility of self-confinement has been suggested again, however, in an objection to the continuum fluid view of the electromagnetic energy used in the virial theorem.[174] In a random neutral plasma the electromagnetic energy is positive. If this portion of the total energy could be negative and sufficient to compensate the kinetic energy, an equilibrium plasmoid is possible. The occurrence of a negative electromagnetic energy in an assembly of charged particles was suggested.

Plasma theory and experiment are as yet hardly sufficient to explain the formation of a long lasting luminous plasmoid under natural conditions. It appears that strong complex external fields would be a requirement for production of any structure if a dense, high-energy plasma is involved.

L. Plasma Models of Ball Lightning

Several plasma investigations were made primarily to study the problem of ball lightning. A plasma of high charge density is involved in theories which deal with the questions of high energy and long existence by suggesting that the ball is formed initially with the necessary energy somehow stored in its contents. An alternative is provided by an external source of energy, the natural electrical activity of the thunderstorm, supplied to the luminous mass continuously over its lifetime. Theories in which the continuous external source is an electromagnetic wave will be discussed in a later section. The substance of which ball lightning is composed has been characterized as a plasma in both types of theories.

The plasma description of ball lightning was anticipated in early suggestions long before the extensive studies of this field in experiment and theory provided the knowledge or the nomenclature now available. In 1905 Carlheim-Gylensköld described ball lightning as a rotating spherical vortex composed of ionized air separated from the cylindrical discharge channel of lightning.[65, 85] The formation of an electron vortex ring by a pulse in ordinary lightning was suggested in 1915. The ring was depicted as rotating rapidly with its electrons ionizing the entrained air by collisions and thus producing a vacuum inside the ball.[581] Similar electron and plasma vortex structures were presented fifty years later.[33, 584]

Other early views of this kind on the composition of ball lightning were not often set forth with sufficient clarity to permit comparison with modern plasma theories, but the special substance of the glowing mass was often described in terms of a high-temperature ionic gas. Thus "fulminating matter" was associated with a temperature of 2500°C, and the spherical envelope was ascribed to electrostatic forces at the surface which confine the interior at high pressure.[310, 311]

The confinement of a completely ionized gas in a sphere solely by the quantum mechanical exchange energy of the electrons provided an estimate of a low equilibrium temperature.[156, 338, 339, 356] Assuming other electrical forces can be ignored in the plasma consisting of free ions and electrons in equal number giving a net neutral charge, the exchange energy is

$$E = -\frac{e^2 n_e h^2}{8\pi mkT}$$

in which h is Planck's constant. Confinement of the thermal kinetic energy $3/2\ kT$ by the exchange energy occurs at the temperature

$$T = \left(\frac{e^2 n_e h^2}{12\pi mk^2}\right)^{1/2}$$

The temperature calculated in this way for an electron density of 2.7×10^{19} electrons/cm^3, equivalent to the molecular concentration of a gas at atmospheric pressure, is 632°K. At greater temperatures the exchange energy would be insufficient to confine the electrons in the structure. This result was interpreted as indicating the possibility of relatively low-temperature spheres which would nevertheless have high charge density. On the other hand, the exchange energy is shown as sufficient to maintain by itself only a low-energy structure. Other parameters of the plasma fluid would outweigh the exchange energy in affecting the properties of the presumed plasmoid, as indicated in the previous section. For example, recombination of the charged particles would limit the existence of such a ball to a small fraction of a second, possibly comparable to the time of luminosity of a lightning stroke. In the initial study,[356] slow recombination was predicted at the "high" temperature calculated. In experimental studies, however, such charge densities have been obtained in the gas phase only at very high temperature, orders of magnitude greater than that given. An electron density of 4.3×10^{18} electrons/cm^3 has been estimated[531] for a lightning stroke at 24,000°K; since no specific change in conditions was suggested in the theory other than the low temperature, it appears that the period of luminosity of such a stroke should be substantially that of the similar matter composing ball lightning. This electron-gas quantum theory of ball lightning, also termed[479] the "crystal theory," is similar to quantum theories of the solid state which have been suggested for application to ball lightning on the basis of the high charge density presumably involved.[382]

The ring current and a spherical structure based on a hydrodynamic vortex were both suggested as ball lightning forms and investigated by magnetohydrodynamic considerations, omitting more detailed charged particle interactions.[434, 449, 473, 474] As indicated previously, external forces are required to obtain an equilibrium structure for such magnetohydrodynamic plasmoids of limited mass in which there is no appreciable gravitational force. In a torus confined with the aid of external pressure, the additional magnetic fields required may be produced by the ring current I_R and a current flowing around the cross section of the torus on its surface I_s

$$\frac{I_R^2}{r^2} = \frac{I_s^2}{R^2 (\ln 8R/r - 1/2)} = \frac{2\pi P}{\ln 8R/r - 3/2}$$

If the internal pressure is negligible compared to the atmospheric pressure and the radius of the ring R is 10 cm and the radius r of the crosssection 1 cm, both currents[474] would be on the order of 10,000 A. The surface current was considered a barrier to penetration of the external gas into the plasmoid. An external magnetic field perpendicular to the plane of the ring could presumably compensate greater internal pressures.[266] The toroidal currents involved in this model resemble those often used in thermonuclear plasma experiments with toroids. The formation in nature of such specific charge flows of the magnitude evidently required, however, is open to question, considering the methods found necessary to produce equivalent currents in the laboratory. The weak constant external magnetic field available for ball lightning, the earth's field (ca. 0.3 gauss), raises similar difficulties in considering the related vortex structure for a spherical region when high energy or long existence are in question.

The absence of strong external fields to assist in the difficult problem of obtaining an equilibrium structure for a plasmoid generated under natural conditions resulted in several theories relying on self-confinement. The conversion of a plasma rod into a plasmoid with the structure of the Hill vortex was suggested as a mode of formation of ball lightning from a segment of ordinary lightning.[86, 87] The cylindrical rod has a strong trapped magnetic field but no electric field. The charged particles revolve around the axis of the rod, and the ions and electrons were described as flowing in opposite directions along the rod. The electric dipole given by the charge separation and the magnetic field cause the respective particles to form an umbrella at each end of the rod. The umbrellas of opposite charge then extend toward each other producing a closed conduction path through the center of the rod, around the outer path, and back to the center once more, giving a spherical outer surface. This resembles the plasma vortex considered earlier[473] with the exception that the external magnetic field used in the hydrodynamic analogy is absent.

The theoretical structure presumably derived on the basis of this model was actually that of a torus with positive ions flowing as a ring current and electron current at the surface around the crosssection of the ring as in previous toroidal structures. Thin rings with the electrons in a very thin outer shell[86] and thick rings more closely approaching a ball-shaped mass[87] were considered. Very high currents of the order of 10^7 A and electron velocities of one-third the speed of light were used in the model. The radiation from the electron current in the form of scattered high-energy electrons and X-rays is thus very high and led to the conclusion that ball lightning is the most concentrated radiation source (5×10^{11} rads/g) produced naturally in the atmosphere. This result derives from the high particle velocities assumed and the limitation of energy loss processes permitted. The accompanying estimate[184] of the duration of ball lightning with an assumed initial energy of 5×10^6 joules and an estimated energy loss of 9.7×10^5 W/sec was 5 sec. Recombination was excluded as a major process, either at high[86] or low[87] internal pressure. This assumption is difficult to accept for the high pressure in view of experimental data, and the low pressure of 1.8×10^{-7} atm used in the alternative model also appears unrealistic for ball lightning.

The formation of a torus with positive ion current was suggested to result from a shock wave disrupting the channel of linear lightning.[187] The pressure in the preliminary lightning channel was taken as 200 atm, approximately 10 times the usual estimate. On cessation of the current flow the channel collapses, generating a strong shock wave as indicated by the acoustic wave. The existence of branches from the main channel of lightning with earlier current decay or the occurrence of a nearby flash of lightning can in this way generate shock waves striking the main channel of the flash involved. The shock wave interrupts current flow, and the magnetic field around the channel collapses at the same time, transforming a portion of the high-energy ion channel into a vortex. The ion velocity is also high in the vortex. The multiple action of several shock waves disrupting a discharge channel in a number of places was suggested to account for the reported sausage-shaped linear discharge shown in Fig. 6. Experimental study of the action of intense shock waves on spark discharges was proposed to test this theory.

A self-confined ball lightning vortex with a toroidal magnetic field and electrons and ions flowing in poloidal currents enclosing the field was considered as the product of a similar transformation of a suddenly isolated segment of lightning.[236] The current flow and magnetic field in this structure are found in exactly reversed roles from those in the vortex based on the hydrodynamic analogy (cf. Fig. 22). The collapse of the magnetic field on recombination of ions and electrons was suggested to accelerate the remaining particles which would maintain the current flow. Collisions of the ions with the surrounding neutral gas would evacuate the interior of the ball giving a pressure gradient determined by the magnetic field strength. The recombination time of the ions and electrons, estimated from published recombination coefficients and an electron density of 2×10^{18} electrons/cm^3, was 10^{-8}–10^{-10} of the duration reported for ball lightning.

The formation of ball lightning as a plasma vortex ring by a pulse from a lightning stroke, especially through a hole in a solid surface struck by the flash, was suggested.[584] A magnetic field was viewed as trapped in the ring, as in the plasmoids produced by strong current pulses in thin wire electrodes. [58] A particle density greater than that of a gas at atmospheric pressure by a factor of several hundred might result from the shock wave of the lightning channel. A high kinetic energy of vorticity up to hundreds of joules may be provided. An estimate of a lifetime of some seconds was made for the magnetohydrodynamic vortex from an approximate rate of radiation of the plasma energy, but recombination rates were not specifically considered.

An alternative formation of ball lightning as a plasmoid was suggested [118] from the "inverted ion magnetron." The confinement according to this model results from the earth's magnetic field and a high radial electric field provided by a linear lightning discharge. The relative strength of the magnetic and electric fields is thus the reverse of that commonly found in laboratory plasma experiments. A spiral wave instability of ions was credited with high energy production in this rotating plasma device. No quantitative description in terms of reasonable fields or the plasma instability was given, and experimental confirmation of the mechanism at atmospheric pressure was not shown. Ball lightning was equated to pinched lightning. This description might introduce an additional difficulty for the theory. Although a strong radial electric field may be preferable for confinement to a self-magnetic field generated by a strong current, data on the radial electric field from a lightning flash, as distinguished from the atmospheric potential causing the discharge and its change during the discharge, seem to be absent. High radial fields from space charge making confinement difficult would presumably be reasonable with the electron current involved in leader processes initiating a lightnig discharge. The occurrence of a confining force from the plasma pinch would be favored during the strong return current during which channel tends to become grossly charge neutral, however. The radial field then depends on the detailed charge distribution across the channel. Any fields would be very short-lived, but confinement from the pinch may increase in significance with neutrality. This is in accordance with the appearance of bead lightning late in long-lasting lightning flashes, as in the event from which Fig. 5 was taken, as well as other reports in the literature.

The formation of ball lightning through the separation of a segment from linear lightning as a result of sausage instability in the plasma channel was suggested in the case of the 1967 observation[128a] on the Onega River in Russia. The ball appeared soon after a flash of ordinary lightning although it was first seen over the river at some distance from the place where the preliminary flash hit on the shore (see Fig. 8). If this observation indicated motion of the fireball away from the shore to its first observed position somewhat closer to the opposite shore and 1.5 m above the water, its motion from that point while in view was uniformly toward the shore struck by the preliminary lightning. The observer compared the bright, yellow-white central portion of the globe, which was 6 to 8 cm across, to the plasma formed in a discharge with a temperature of 14,000°K. A temperature of 13,900°K was calculated from the relationship for the half-life of ion recombination in the ball

$$t = 1.01 \times 10^{-20} T^{-1/4} \exp(31.5 + 8.8 \times 10^{-4} T + 7.06 \times 10^4 T^{-1})$$

in which the half-life t in minutes is given as a function of the plasma temperature T. For the half-life a period of 1.35 min was evidently used. This value is actually the whole life of the ball estimated with the assumption that it originated in the initial linear flash. A recombination coefficient of 10^{-19} cm^3/sec was used. The plasma in the ball lightning was assumed for this calculation to be positive ions and electrons, negative ions being formed only to small extent. The initial ion concentration was taken as the equilibrium amount for the temperature, with recombination proceeding throughout the life of the ball with no regular decrease in the temperature. Experimental plasma studies have shown repeatedly that the lifetime of a normal plasmoid with the temperature indicated, if set free in air without a continuing supply of energy from the outside, would be several orders of magnitude shorter than 1.35 min. It thus appears that the model of recombination used in this calculation is entirely inadequate. A stabilizing effect of recombination against temperature changes was suggested. A decrease in temperature in the central region of the ball would result in an accompanying increase in the rate of recombination restoring the higher steady-state temperature. A rise in the central temperature would both decrease recombination and increase heat conduction out through the surface of the ball, causing the temperature to fall again to the equilibrium condition with some loss in energy. The initial ionization was estimated from the temperature at 22%, corresponding to an ion concentration of 1.2×10^{17} ions/cm^3 in the high-temperature ball lightning gas of presumably lower-than-atmospheric density. Using a mean ionization energy of 14.5 eV the total energy in the ions is 530 joules in the lightning volume of 1450 cm^3. The theory of temperature oscillation following recombination to compensate for a loss of energy was extended to explain repeated pulsations which might occur, since appreciable energy would be available in the estimated ionization. Radiation emitted by the ball would pulsate along with the oscillations in temperature and energy release from the recombination. This concept of departure then return to a stable equilibrium temperature requires recombination to lag behind a relatively large energy loss. If the entire energy represented by the initial ionization were released in 10^{-4} sec, the explosive power would be approximately equal to that released by 100 g of nitroglycerin. The initial processes in formation of the lightning ball structure were suggested by Dmitriev as departure of electrons from the central high-temperature region followed by their attachment to oxygen molecules, thus forming a shell of negative oxygen ions. The shell must be stable and provide a barrier to diffusion of charge, otherwise the ion mobility would result in dissipation of the plasma ball in less than 0.01 sec. The layer of low-temperature oxygen ions may trap electrons in the center of the ball, and bombardment by the electrons can produce some excitation and luminosity in the shell. The motion of the sphere was attributed in part to a small variable positive electric charge of approximately 3 microcoulombs generated by the discharge of a microampere corona current. According to this theory the continued existence of the ball despite violent sparks produced on collisions with the trees was ascribed to the outer stable envelope of negative oxygen ions (O_2^-, O^-) formed by capture of electrons. Contact with grounded metal would presumably neutralize the stabilizing shell and destroy the structure.

The existence of ball lightning within structures which shield against external electric fields and its penetration through dielectric substances was cited as evidence that an external source of energy cannot maintain the plasma fireballs. The plasma spheres were depicted as intense, high-frequency microwave fields within a cavity formed by highly ionized, conducting spherical walls carrying large surface currents.[119] The energy is stored in the radiation field rather than in the plasma alone. Such resonant microwave cavities might be formed by a small loop in a lightning channel or a strong current in a lightning conductor. The long lifetimes require minor dissipation of the energy present, a requirement which was met by use of a high electromagnetic frequency, much greater than the collision frequency of electrons and molecules and confined wihin the spherical volume by the highly reflective walls. Maintenance of the plasma with low dissipation of energy can only occur at reduced pressure, and the external pressure exerted on the globe by the atmosphere is balanced by the radiation pressure of the internal high frequency. A single pure resonant mode of the electromagnetic field would not give a uniform radiation pressure over the surface, so several coupled modes would be needed. In such a structure a high frequency of 1.6×10^{10} Hz with a field of 3×10^8 V/m would provide the desired radiation pressure, 10^5 newtons/m^2, while maintaining a density of 6×10^{10} electrons/cm^3. The assumption that the energy in the internal field is largely determined by the need to balance atmospheric pressure leads to a content of 400 joules in a sphere of 10 cm radius and 10^4 joules one in of 30 cm radius. Plasmoids generated by microwave frequencies approximately one-tenth the frequency and 10^{-5} the field strength indicated in this theory have been reported, as discussed in the following section.[404, 405] These plasmoids were maintained by the external radiation for periods of 10–20 sec, indicating absorption of the energy provided through the walls, but on cutoff of the energy supplied the visible glow continued for 0.5 sec or more. The lifetime is in reasonable agreement with expectation from the radiation–field energy storage described when the altered values of the parameters and the reduced charge number in the supposed wall are considered. The authors of the experimental work, on the other hand, ascribed the continuing glow to slow decay of energy-storing metastable states of the light-radiating gases. The excitation of the extremely high-electric-field microwaves described in this resonant cavity theory presents a difficulty even for the great electric storms of nature.

An unusual theory of ball lightning proposed a relatively small toroidal core of oxygen and nitrogen atoms in a dense superconducting metallic state surrounded by a plasma.[503] The magnetic field from currents in the core binds the plasma in the structure. The compression of a linear lightning channel generates the metallic-state core. The discharge channel is at first subject to the usual magnetofluid mechanic pinch effect. The resistance of the channel increases in the region of the nodes which result. The current flow then occurs largely in the outer layer of the channel giving strong surface electrodynamic compression with superhigh pressure and density at relatively low temperature. The electrodynamic pinch produces a metallic state from the atoms of the air gases. The region of the lightning discharge then is twisted into a loop which is separated from the remainder of the channel. An energy content of 1400 kcal/g was estimated for the core material. The outer plasma around the core is at a high temperature and radiates the light associated with ball lightning.

The explanation of ball lightning as a plasmoid encounters some fundamental difficulties. First, there is the problem of finding an equilibrium structure and maintaining it against the dynamic plasma instabilities. An external force such as the atmospheric pressure appears necessary to supplement self-fields in the plasmoid which cannot by themselves maintain a closed system with the required conservation of energy according to conditions imposed by the virial theorem. The upper limit placed on the energy of the plasmoid if atmospheric pressure is the sole external force assisting confinement is much lower than several estimates[405] of high energy based on reports of heat release or explosion of ball lightning,

$$PV \geq U + E + M$$

which limits the total energy contained to approximately 100 joules/liter. The parameters given by the virial theorem, however, do not include additional "chemical" energy which may be present in ionization, excitation, or chemical reaction energies of the plasma fluid. Thermal loss, up to the maximum permitted by the virial theorem at the surface of the plasmoid is too rapid for continued existence of the mass beyond a few milliseconds without some insulation. The inability to account for confinement in a plasma of the energies often considered in some kinetic, thermal, or electromagnetic form for the lifetimes involved has led to the theories in which ball lightning is maintained by an external source of energy.[151] The formation of ball lightning as a spherical dc electrical discharge has already been discussed. In the following section natural electromagnetic waves are considered as the source for such a plasmoid.

The observation of luminous balls formed in air by high-power electrical discharges in metal holds some interest for this problem.[237, 477] Such fireballs have been described as plasmoids, and the slow recombination indicated by lifetimes on the order of a second has been ascribed to configurational energy.[477] No specific configuration capable of retarding recombination in this way is known. The reports of these globes indicate, however, that long-lived luminous spheres, possibly incandescent gas vortexes if not unknown plasma structures, are capable of existence.

M. Formation of Ball Lightning by Natural Electromagnetic Radiation

The generation of ball lightning as a natural plasmoid by an electromagnetic wave was proposed by P. L. Kapitsa in 1955 in a paper which greatly influenced later work in the field.[241] Kapitsa concluded that a continuous external source of energy is necessary to account for the long period of luminosity observed in cases of ball lightning. The maximum energy which could be stored in such a ball when it is formed is inadequate to provide continuing luminosity for the period involved. An estimate in support of this argument was obtained from the radiation time of the fireball from a nuclear explosion. Such a cloud should contain the largest possible store of energy and should thus provide an estimate of the longest time for which a ball can glow using only internal energy. The emission of light for the order of 10 sec from a nuclear cloud with a diameter of 150 m was used as a basis for the estimate of the life of the small globes of ball lightning. The quantity of energy in such a cloud is proportional to the volume and thus to the cube of some linear dimension of the cloud, and the light radiated is proportional to the area of the surface which is a function of the square of a linear dimension. In this way Kapitsa concluded that the radiation time from a ball is proportional to a linear dimension such as the diameter. A ball 10 cm in diameter can glow for a maximum of 0.01 sec if related in this way to the nuclear cloud. The duration of ball lightning is much longer than this, indicating that energy must be fed into the ball to continue its luminosity for a longer period.

Very small plasma fireballs approximately 10^{-4} cm^3 in volume generated by a focused laser give additional information on the lifetime which may be expected.[20] When a glowing ball was produced with initial energy of 1 joule provided by the light source, a duration of some 10 μsec was observed, in agreement with the approximate relationship $t = kE^{1/2}$, in which the lifetime t is determined by the initial energy E. This approximation is applicable to very large as well as very small fireballs. The radiation from the luminous could of a 1 kton nuclear explosion lasts for the order of 10 sec [i.e., $k \sim 10$ sec/kton$^{1/2}$]. Using the estimates for high-energy ball lightning,[184, 512] 11×10^6 joules (3.3×10^{-6} kton), the lifetime of the glowing sphere should be on the order of 0.01 sec, in accordance with the estimate by Kapitsa based on the size of ball lightning. Again, the existence of ball lightning appears to exceed by far the period which is reasonable for even the highest initial energies suggested.

Kapitsa suggested that a natural radio wave is the external source of energy required to account for the unexpectedly long lifetime of ball lightning. The diameter of the spherical ionic plasma which is ball lightning is determined by the resonance of the ball with the electromagnetic oscillations of the external wave at which absorption of the external power is most effective. This resonance occurs for a sphere when the wavelength of the external radiation is 3.65 times the diameter of the sphere.

The region where the ball lightning will form begins with a low concentration of ionized gas formed by a preliminary lightning stroke or other electrical activity in the storm. The resonance of a smaller region with the external wave occurs as a result of the lower ionization level, and the energy from the radio wave is effectively absorbed as the ionization increases, producing a corresponding increase in the volume of the sphere until the equilibrium diameter is obtained. Compensating processes maintain the equilibrium size of the ball. For example, an increase of power in the external radiation raises the temperature of the plasma, and the volume of the ball increases accordingly. The larger globe is no longer in resonance with the electromagnetic wave which results in a decrease in the absorption of power and a return to the equilibrium dimension. The most common ball lightning dimensions noted in observations would be associated with electromagnetic wavelengths of 35–100 cm.

The formation of standing waves by reflection of plane polarized radio waves from the surface of the earth was suggested as a mechanism by which the electric field of the incident wave is increased and the position of the ball formed is determined with respect to the ground. The antinode of the standing wave produced by reflection would have an electric field double that of the initial radiation giving a favorable location for ionization. The fixed height of the stationary antinode in the electromagnetic field prevents the ball formed from rising like the nuclear fireball. According to this theory the passage of ball lightning into buildings involves the transmission of the short radio waves through waveguides provided, for example, by chimneys. Kapitsa considered among the problems raised by this theory the question of the then unreported electromagnetic waves required from the electrical activity of a storm. He suggested that oscillation of an ionized gas either near the cloud or the ground is the source of the radiation.

The generation of ball lightning as the electrical discharge given by a standing wave had been proposed by earlier investigators. De Jans[230] credited Lodge[288] with the view that a standing electrical wave was produced in an underground metal conductor struck by lightning. The peaks in the electric potential of the wave form ball lightning as a brush discharge seen above ground. The term standing wave is used explicitly in de Jans' review of 1912 and not in the earlier work of 1892 to which he referred. A similar process was invoked in 1930 to explain the formation of ball lightning, previously described, in a closed room.[303] A globe the size and color of an orange appeared at the metal door handle inside the room as thunder sounded. There was a loud explosion as the ball disappeared without having moved. The observer suggested that a standing wave may have been formed in the telephone lead wire along two walls of the room as the result of a high-frequency current induced in the telephone wires by a lightning discharge. Kapitsa also specifically noted the problem of generation of ball lightning in enclosed spaces, particularly inside airplanes which are almost totally enclosed by metal, and concluded that the electromagnetic wave theory effectively accounts for this phenomenon. The appearance of very bright balls of different colors from a stove in a building near the summit of a mountain was ascribed also in 1930 to the probable agency of a stationary wave.[429] Lightning was noted in the distance although no rain or hail was falling from the dark cloud near by. A more recent presentation of the standing wave theory, lately credited with priority,[479] was given in connection with the appearance of ball lightning in adobe huts in a Mexican desert.[95] Formation of standing electromagnetic waves excited by a lightning stroke in the adobe room acting as an electromagnetic cavity was proposed. Kapitsa's work, however, undoubtedly provided the basis for the extensive consideration of standing-wave theories of ball lightning which followed.

Several aspects of plasmoids and electromagnetic waves have been investigated experimentally and theoretically. The results have provided support of some parts of the theory and additional questions on other parts. Experimental plasmoids in high-frequency electrical fields increase in size with an increase in the power provided by the field.[141, 201] The constant size and brightness of most of the lightning globes reported in a survey[420] were cited in support of an external source of energy for the ball, since expenditure of an initial store of energy over the life of the ball should be accompanied by marked changes such as a decrease in size. On the other hand, 15% of the reports in this survey indicated a change had occurred, and shrinking of the ball or marked change in its color is common.

Investigation of the confinement of charged particles by a polarized standing electromagnetic wave by means of the Mathieu equations in an extension of Kapitsa's theory showed that particles are bound near the electric node by this type of force.[562] The lack of solutions for the Mathieu equations at the antinodes of the electric field indicates the absence of this confinement. The binding obtained in this way from the standing fields is two-dimensional. The motion of particles along the wave is not affected, but the particles are confined against transverse motion when the field is produced by reflection as suggested by Kapitsa. The usual horizontal motion of ball lightning thus requires lateral motion of the confining wave along the ground. This harmonic field confinement evidently differs from that conceived by Kapitsa, who suggested the intensified field at the electric antinode given by reflection to favor formation of plasma by ionization with particles of higher energy in a region at a fixed distance from the earth. Harmonic binding of the charged particles according to the Mathieu equations requires interaction of the particles in the plasmoid with the external field. This can occur only at low charge densities; for penetration of the field is prevented at moderate plasma densities, resulting either in distortion of the field inside the plasmoid or absorption of the external energy only at its surface. An additional electromagnetic force resulting from motion of the charged particles in the electric field with the magnetic component of the electromagnetic wave may also conflict with the harmonic confinement[167] although this difficulty disappears at resonance of the plasma with the external wave. The localization of the ball at the electric field nodes where the field gradient is a minimum was supported in these studies.[167, 520] The problem involved in the shift from the antinode where ionization is favored to the node where a low impedance plasmoid is best located has been pointed out[520] with the accompanying difficulty of retention of the standing wave fields. The formation of characteristic discharges at both nodes and antinodes of high-frequency radiation has been reported.[27, 479]

Use of the radiation pressure from an external electromagnetic wave to prevent a ball 10 cm in diameter with a temperature of several thousand degrees from rising with its buoyancy of approximately 600 dynes appears difficult. Twenty dynes, less than 10% of the required balancing force, was estimated as available from 40 kW, half of which is utilized in maintaining the ball.[520] The buoyancy of the approximately 13-cm-diameter ball lightning with estimated temperature of 14,000°K reported over the Onega River was evaluated as about 1700 dynes.[128] The force preventing the ball from rising was ascribed to electrostatic attraction of the earth or grounded objects for the excess charge of 2.8 microcoulombs present with some variation on the ball as a result of escape of electrons in a corona current. General motion of the globe was also considered a result of the unstable electrostatic equilibrium in which it would be found. The possibility that its position would remain constant even in absence of a charge was presented as reasonable on grounds that the estimated buoyancy is very small. An estimate of the power required to maintain such a ball was obtained by assuming that the power dissipation is the same as that of an electric arc with the same surface area. A requirement of 18 kW was found for the 10-cm ball using a power dissipation of approximately 60 W/cm^2 in the electric arc. Radiation providing approximately five times this quantity of power, well-focused and with a frequency constant to perhaps 2%, was considered necessary on this basis for the generation of ball lightning in nature, and the formation in nature of a 40-cm wave providing this power was described as highly probable.[520]

Consideration of the confinement of charged particles by harmonic binding was extended to the fields given by three orthogonal standing waves. Full three-dimensional containment of the charges particles is possible in such a field, in contrast to the fields studied previously which exert no force on the particles along one space coordinate.[475] The equations for the three-dimensional system were too complex for a general analytic solution, but numerical computer solutions were obtained. The results indicated that 2.5 MW of power in a frequency of 1.6×10^9 Hz could confine an electron with an energy on the order of 50 kV near a pure electron cloud containing 10^{16} electrons/m^3, considering the confinement region as a cavity with wall conductivity of 3×10^6 mhos/m (Q power factor of 50,000). The formation of three such waves in the necessary geometric relationship in nature seems unlikely. An alternate model of a standing wave field considered is that formed by a pulsed narrow-band, polarized radio wave in place of the continuous radiation used in earlier theories.[476, 479] A pulsed interference field is produced by reflection of such a wave, resembling a process in radar telescopy.

Certain ball lightning occurrences seem more readily explained with the theory of an external source of energy in the form of high-frequency radio waves. The possible role of a chimney as a waveguide suggested by Kapitsa is indicated frequently. Three witnesses saw a glowing ball the size of a bowling ball form over the hearth at the moment of a strong lightning stroke outdoors. The ball radiated a blue light, compared to that of an alcohol flame.[413] It was rotating, and as it rolled on the hearth and fell on the floor the witnesses fled. One of the observers happened to be looking directly at the fireplace when the ball appeared and concluded that the ball did not come from outside. In another incident a crackling fireball tne size of an apple appeared in a fireplace following an intense explosion when a fire was made two hours after a strong flash of lightning struck a tree outside.[229] The storm was long past, and the sky was clear. The incident of a tub of water heated to boiling by ball lightning has been cited as a contradiction to the theory of external high-frequency radiation as the source of power,[277] because radiation of this type does not propagate in water. The most widely accepted estimate of the high energy content of ball lightning was based on this example. A significant observation supporting the presence of high-frequency radiation, however, was reported in 1911 (page 123 of reference 65). An occurrence of ball lightning was credited with causing the filament of an electric light to glow. The light from the bulb was very bright, and it went out suddenly when the ball lightning disappeared. The theories involving an external radio wave supporting ball lightning have been alternately rejected as unnecessary to account for the energy content[213] or accepted as providing a clear mechanism for specific occurrences, such as the generation of ball lightning near the nose of aircraft.[33, 257]

Experimental studies with high-frequency radiation have shown the formation of luminous discharge regions under conditions analogous to those suggested in the standing-wave theory of ball lightning. Pulsed radar beams were focused with the aid of parabolic reflectors to obtain the glowing masses in evacuated glass bulbs below atmospheric pressure.[170, 195, 435] A barrel-shaped discharge with a diameter of approximately 4 cm was generated[170] by a 3-cm wave at a pressure of 10 mm Hg. The diameter decreased to 1 cm as the pressure was increased to 40 mm Hg. At the lower pressures the discharge was red–violet; at higher pressure it was more intense and violet in color.[435] The shape of the luminous region changed with the position of the glass bulb and with the pressure. Irregular forms with striations were present at 3–20 mm Hg. At lower and higher pressures one or two sausage-shaped discharges appeared in the bulb along the axis of the reflector. Increasing the pressure in the bulb extinguished these discharges at less than 50 mm Hg pressure. An average of approximately 50 W was provided by the pulsed beam. The wall of the chamber remained cool except when the discharge region actually touched the glass. When in contact with the bright mass the wall became too hot to touch in 10 sec. A bright flame-like discharge was produced in air at atmospheric pressure by a high-frequency oscillator tuned to a coupled half-wavelength wire.[478] Breakdown was reported to occur only at the antinode of the electric field, and the geometrical shapes observed at lower pressures in the fields produced by reflection were absent. The 10-kW level of the source was inadequate to provide the 10^6 V fields indicated as necessary for ionization in the standing wave theory with the ball located at the field node,[562] but the discharge appeared to be the high-voltage type.

Very bright, almost spherical plasmoids approximately 2.5 cm in diameter were generated in a cylindrical quartz tube encircled by an induction coil supplied by a 10-MHz radiation source.[334] In contrast to the prevalent opinion that the generation of high-frequency discharges at atmospheric pressure or higher would require high power on the order of 10 kW or more, the plasmoids could be maintained in argon and krypton by 600 W. The discharges were initiated at low pressure, and the pressure was gradually increased while power was supplied. The most effective operation of the power supply was obtained by empirical adjustments of the generator during the discharge, which were important in formation of the plasmoids. The power utilized in maintaining the plasmoid was determined by the difference between the total generator power and the power dissipated by radiation from the anodes of the oscillator tubes measured with a thermopile. Impurity lines in the radiation from the plasma were very weak. The most intense lines from argon were neutral lines, and at higher pressures the intensity of the visible continuum increased greatly. The temperature of the high-pressure argon plasmoid was estimated as 9000°K with an electron density of approximately 2×10^{16} electrons/cm^3. Rotation of the gas in the discharge by means of a propellor at the bottom of the experimental chamber, suggested to provide stability for the plasmoid in accordance with some views of the requirements for stability of a plasma sphere and with observations of rapidly rotating ball lightning,[482] proved ineffective. In a 60-Hz ac discharge between electrodes, rotational motion of the gas was required for a stable discharge.[552] The effect of the weak vortex, which also reduced rf noise from the discharge, was compared to the luminous electrical phenomena in a tornado funnel. The plasmoid in the induced electrodeless discharge, on the other hand, was stable at high pressure without the added motion, and rotation of the propeller produced a twisting plasma pinch which was extinguished by high pressure.

Long-lived glowing masses were generated in air at atmospheric pressure by a high frequency discharge[404]. The fiery regions, actually bent or snakelike rods with 5–7 cm diameter, continued to glow 0.5–1 sec after the power was turned off. The discharges were formed by a 75-MHz wave from metal electrodes inside a cubical chamber 2.5 m on a side with aluminum walls. The electric field was estimated at 1000 V/cm in the TM_{110} resonating cavity. When power was continued the free-floating ball traveled to the wall in 10 or 20 sec. Occasionally it exploded. Study of these high-frequency discharges was extended by Powell and Finklestein to spectroscopy of the species responsible for the light radiated, using a 30 kW source of the frequency in 15-cm-diameter glass tubing.[405] The discharge was formed between electrodes. When the power was discontinued, the plasma detached itself and rose to the top of the cylinder where it remained until its luminosity ceased. In the glass chamber the glow continued 0.5–1 sec after power was turned off.

The same discharges generated in the open without confinement by the glass remained visible for 0.2–0.4 sec. These fireballs formed at pressures of 0.5–3 atm, the latter being the highest pressure studied. Below 0.5 atm the discharges were more like conventional glow discharges and disappeared quickly when the power was shut off. The glowing masses with long lives occurred in nitrogen, oxygen, mixtures of nitrogen and oxygen in ratios from 20:1 to 1:2 (including the one corresponding to air), and nitrous oxide. In argon and carbon dioxide arc-like discharges were formed, and these glowed only a few milliseconds after power was ceased. The radiation in nitrogen was bluish and relatively weak. In oxygen it was white and very bright, so intense that it was difficult to look at. In ordinary air (20% oxygen) the discharge was yellow-white and of medium brightness. The fireball in air became more yellow in time as nitrogen dioxide accumulated in the discharge gas. The largest balls, sometimes over 50 cm in diameter, formed in nitrous oxide and were markedly orange in color. These lasted up to 2 sec after power stopped. Various electrode metals were used successfully in producing the discharges, including platinum, gold, silver, copper, zinc, cadmium, tin, aluminum, carbon, and tungsten. Lead electrodes or mercury-coated electrodes were unsuccessful. The investigators concluded that, in general, less easily vaporized electrodes gave the longest-lived discharges. The temperature of the fireballs was indicated as 2000–2500°K by the resistance of 75-μ tungsten wires. This temperature is far below the temperature at which air normally glows appreciably (ca. 4000°K), and accounting for the observed luminosity thus becomes a difficult problem. The total radiation in the visible and short infrared from 0.4- to 1.1-μ wavelength was measured. The brightest discharge, that in oxygen, emitted up to 160 W. The considerably dimmer glow in air radiated approximately one-tenth of this in the visible region between 0.4 to 0.72 μ and even less in the infrared, from 0.72 to 1.1 μ. A concentration of 10^{15}–10^{16} per cubic centimeter of excited species with 3–6 eV in excitation energy would provide the observed luminosity. Approximately ten times this was estimated to provide the energy lost to the chamber wall. An electron concentration 3×10^{12} electrons/cm^3 was measured in nitrogen 100 msec after power shutoff. The visible light from the discharges in all of the gases proved to be radiation from impurities rather than spectra characteristic of the gas in the chamber. Radiation from substances entering the discharge from electrodes or the wall of the experimental chamber is often reported[132] in plasma experiments. In nitrogen all of the visible glow from the discharge came from metallic species from the electrodes. Organic impurities in the apparatus also contributed CN species sometimes noted in the spectra. The radiation from the discharge in air was quite different from that of nitrogen and essentially the same as pure oxygen. The visible light was largely the carbon dioxide continuum from the carbon monoxide–oxygen atom reaction with several of the electrode-metal lines also present. At wavelengths below the visible spectra the Schumann–Runge bands of the oxygen molecule were noted at moderate intensity and the OH(0,0) band of strong intensity but not significant in the visible glow in view of its wavelength of 3064 Å. The color produced in the discharge also indicated the formation of nitrogen dioxide, but its spectrum could not be distinguished in the presence of the carbon dioxide radiation. The low measured temperature of the discharge and the low probable concentration of carbon dioxide (less than 1%) require a nonequilibrium excitation much greater than the actual kinetic temperature to account for the strong carbon dioxide emission. Metastable electronic levels of molecular nitrogen or oxygen which transfer stored energy to the radiating species responsible for the visible luminosity were suggested.[405] The $A\,^3\Sigma^+_u$, $\omega\,^1\Delta_u$, and $^3\Delta_u$ states of molecular nitrogen were proposed to account for the long-lived radiation in nitrogen alone. In air the collisional deexcitation of such nitrogen levels would presumably be very rapid, and considering the greater brightness of oxygen post-discharge luminosity the $b\,^1\Sigma^+_g$ and $a\,^1\Delta_g$ molecular oxygen states were proposed to account for the long-lived glow in air. Although the radiative lifetime of these species ranges from 1 to 10 sec, collisional deexcitation in air at atmospheric pressure and 2000°K, according to available rate coefficients, would be so rapid as to effectively remove the energetic species in less than a microsecond, with the possible exception of the $a\,^1\Delta_g$ state of molecular oxygen. In view of the essential role of carbon dioxide, metals, or similar species with strong visible emission in this model for the luminosity of ball lightning, formation of the fireball according to the theory depends on generation of a radiating mixture containing these substances. Such a mixture could form as the result of a preliminary linear lightning stroke to a metallic or wooden object. The observed radiating species were in general substances extraneous to air, with the possible exception of nitrogen dioxide. Thus the relatively low temperature luminosity differs from the high temperature radiation of the normal lightning flash.

The estimate of 1000 V/cm for the electric field present in these rf discharges, which is larger by several orders of magnitude than rf fields observed in nature, led the investigators to eliminate the possibility that natural rf radiation is the source of ball lightning. Thunderstorm dc fields of this magnitude are reasonable, on the other hand, and such fields were suggested instead as the cause of ball lightning. The glowing sphere is in this case a positive-ion globe produced in a dc discharge. The continued existence of the rf discharges for 0.5–1 sec after power is off was, however, considered applicable to the dc discharges as well. Continuation of the discharge beyond this period for the several seconds often reported as the life of ball lightning utilizes current from the electric field of reversed polarity remaining after a preliminary lightning discharge. The reversed field polarity maintains the position of the positive globe against buoyancy. Thus rf experiments indicating the possibility of a long-lived luminous ball after external power has ceased were used as a basis for a dc theory of ball lightning in which continuing power is supplied by a thunderstorm field.

The reported lifetimes of the reversed polarity field following lightning are not longer than the period of luminosity after power is off in the rf discharges. The use of the rf glow time for dc discharges is reasonable in view of the processes considered to produce the metastable species which are credited with the long-lived storage of internal energy. There is a large difference, however, between the period of luminosity from the rf discharges ($\sim$1 sec) and the visible lifetime estimated from experimental data for a 10-cm sphere with maximum initial energy from a nuclear explosion or a laser-generated plasmoid (0.01 sec). The exceptions to the short visible period of a high-temperature plasma, lightning, rf or dc discharges, and laser plasmoids are found in the luminous objects from dc discharges involving vaporized metal discussed previously and in these rf discharges in which extraneous substances including metal from the electrodes are present. The metastable species suggested by Powell and Finklestein, although as yet unobserved directly in the rf discharges, provide a possible explanation of their markedly longer visible luminosity. The short-lived discharges may well involve excitation levels of higher energy in accordance with the greater initial temperature and energy in the specific examples cited. The long-lived metastable states, on the other hand, evidently require specific excitation processes under moderate conditions.

Radio wave plasmoids known only at low pressures have been reported with dark rings surrounding a central luminous zone.[525] The dark-ring plasmoids formed only when concentrations of both neutrals and ions were low, with electrons at a concentration of approximately 10^7 electrons/cm^3. Low electric fields are thus sufficient. Study of these discharges in purified gases showed that oxygen, sometimes present only as an impurity, was necessary to obtain the dark-ring plasmoid. The characteristic color of the discharge in oxygen was that of the yellow–green 5632 Å line of the molecular oxygen ion O_2^+. Comparison of these plasmoids with the ball lightning with a bright yellow-white central area surrounded by a dark violet layer and an outer light blue corona, which was observed at close range from the bank of a river in Russia,[128] is of interest.

Information on the occurrence of the high-frequency radio wave required in nature for the generation of ball lightning according to the standing wave theories is sparse. Radiation with a wavelength of 73 cm and approximate frequency of 410 MHz produces ball lightning with the often reported diameter of 20 cm if the plasma resonance suggested in Kapitsa's theory is effective. Kapitsa himself proposed that the required microwaves might be generated in thunder clouds and either radiated as a focused beam by the clouds to the ball or propagated along the ionized channel from a lightning discharge.[243] The presence of a megahertz frequency in a lightning column was suggested on the basis of a rough estimate obtained from copper surfaces hit by lightning.[214] Closely spaced concentric rings were seen around the fulgamites raised in the surface. From the spaces between rings and estimates of the properties of the liquid copper formed by the stroke, the velocity of the ripples was obtained assuming they were caused by resonant acoustic oscillations. A radio acoustic wave was presented as the possible source of the approximate megahertz frequency.

Observation of microdischarges at 30- and 50-MHz frequencies were reported from cumulus clouds, although laboratory studies with charged water drops failed to reproduce the electromagnetic radiation observed experimentally with mercury drops, which act as quarter-wave antennae in radio emission.[451] The parameters involved in generation of high-energy ball lightning with electromagnetic radiation produced by rain drops in this way were considered, assuming that all of the electrostatic charge on the rain drops is converted to the electromagnetic radiation.[8, 9] A total energy of 5×10^6 joules was assumed necessary. Using as drop parameters in a medium rain an average radius of 0.05 cm and charge of 0.1 esu with a concentration of 400 drops/m^3, opposite charges being present in equal quantity, 2.5×10^{12} m^3 of rain would be required to obtain the total energy. A higher maximum charge approximately 10^7 times as great was considered possible on the basis of the maximum charge attainable by solid particles. A lower average charge of 100 esu/drop, which would require 10^6 m^3 of rain to produce 5×10^6 joules, was considered reasonable. The possible activity of lightning in producing high concentrations of strongly charged drops was discussed.

Several difficulties in this explanation of the source of electromagnetic energy for ball lightning were pointed out in a critical discussion[291] which resulted in retraction of the theory by the initial author.[9] The highest charge on a drop suggested for consideration, 10^6 esu on a drop with a radius of 0.05 cm, was indicated as far greater than the maximum limits imposed by field emission and by the surface tension of the drop liquid. Field emission would reduce the charge on such a drop by the ejection of an ion or an electron in the electrostatic field given by the charge itself, or disruption of the drop would occur by the inability of the surface tension to withstand the force of such a great charge. Water drops of slightly larger radius were disrupted by a charge only 1% of this. In addition, the energy intensity needed for formation of the fireball presumably requires a well focused discharge in 10^{-8} sec by collision of all the drops in a very large region. The energy might normally be expected to radiate uniformly in all directions from the cloud. In acknowledging the evident failure of this theory to provide a reasonable mechanism by which the energy for ball lightning might be obtained, the author suggested that a low power density, only tens of watts per cubic centimeter of the discharge is sufficient to sustain a high-frequency discharge in air. The droplets could thus discharge their energy in a much longer time than estimated in the objection although the difficulty with focusing in the energy radiated is not overcome in this way. It may be noted that the drop charge used in the moderate estimate was far below the maximum limit indicated, but a large cloud volume would then be required to provide the desired energy. On the other hand, the possibility of a source of electromagnetic radiation of this type in some cloud process is certainly indicated, and consideration of generating mechanisms, while difficult, appears necessary. Until a specific process has been examined, the reasonable expectation that radiation is isotropic is not a conclusive difficulty.

The electromagnetic radiation in storms, especially that accompanying lightning, has been observed in a broad range of frequencies. These atmospherics were recorded simultaneously at frequencies from 6 kHz to 450 MHz in a number of storms.[221] The measurements were made at distances of 5 to 20 km from the storms. The amplitudes varied inversely with frequency from approximately 10^4 μV-sec/m at 10 kHz down to 1 μV-sec/m at 10^5 kHz whereas the field required for breakdown in air to establish the lightning -ischarge is of the order of 10^6 V/m.

Frequencies from 100 to 4000 MHz with vertically polarized electric field were observed during lightning discharges, many within 3 km of the receivers.[261] In some measurements the distance of the lightning was determined by the time interval between arrival of 6 kHz–2 MHz radiation and the sound of thunder at the laboratory site, and the electromagnetic signal was recorded beginning 5 msec after the lower frequency triggering radiation. Photographs of oscilloscope pulses were obtained at three frequencies: 140 pictures of 100-MHz radiation, 40 of 400-MHz waves, and seven at 800 MHz. Signals at 1300 MHz were seen on the oscilloscopes about ten times. The 100-MHz signals lasted 0.2–5 msec. The 400-MHz signals usually lasted 0.1–0.2 msec, although in a few cases the radiation continued up to 2–2.5 msec in the form of a succession of 50–100 μsec pulses. The 800- and 1300-MHz radiation usually lasted 50–100 μsec. The maximum voltages observed at 100 MHz decreased approximately inversely with the distance from the lightning, from 150 μV at 2 km to roughly 40 μV at slightly more than 7 km. The power of the 100 MHz signals was usually 10^{-11}–10^{-10} W in a 100-kHz band, and the 400- to 1300-MHz signals carried 1–5 $\times$ 10^{-10} W in a 1-MHz band. The power indicated at the receivers was converted to comparable values by use of the measured lightning distance and the estimated effective antenna areas. Studies of lower frequencies from 5 kHz to 500 MHz indicated a decrease in power approximately proportional to the reciprocal of the square of the frequency, $1/f^2$ (see also Ref. 250). In this work on higher frequencies, which indicated radiation up to 2000–3000 MHz, the power did not follow the $1/f^2$ noise dependence. The possibility of another maximum in the power radiated at higher frequencies was suggested.[261] Kapitsa pointed out that these results show narrow-band frequencies near appropriate wavelengths for the generation of ball lightning according to the standing wave theory although the radiation is weak and short-lived.[243] He suggested that such radiation might in rare cases be sufficiently strong and long-lasting to provide the continuing energy required to produce ball lightning.

The emission of radiation with frequencies from 400 to 1000 MHz was observed in different stages of lightning.[69] The high-frequency radiation was noted with the stepped leader, the dart leader, and occasionally from the return stroke which sometimes exhibited a delayed emission of 60–100 μsec. The electromagnetic radiation is evidently emitted in connection with a breakdown process as indicated by the radiation pulses observed when large electric field changes occur. The origin of the radiation in or near the cloud was suggested by the delayed emission sometimes observed.

The characteristic sounds associated with many observations of ball lightning may be related to plasma radiation over a broad range including acoustic frequencies and audible sound, sometimes designated electromagnetic sound, occasionally noted with high-frequency radio waves.

The theory of ball lightning as a plasmoid generated by natural electromagnetic, high-frequency radiation is under the most active consideration. Several characteristics displayed by the natural fireballs which are unexplained in other theories can be accounted for in this model of the glowing spheres as radio frequency plasmoids, including their formation in closed rooms, their entrance into buildings through chimneys, their association with other electrical thunderstorm discharges, and their long life. The high frequencies required in nature for production of such plasmoids have been observed in connection with thunderstorm activity but of very low intensity, with electric fields, duration, and power inadequate by several orders of magnitude to provide the theoretical requirements. The actual source of the radiation observed in thunderstorms is unknown, as are the specific processes responsible for the light from the natural balls. Experimental investigation of plasmoids formed by electromagnetic radiation has shown that such luminous spheres can be generated by high-frequency radio waves at atmospheric pressure and maintained with modest power. A more favorable aspect has thus been provided for some of the most difficult problems encountered by this relatively recent theory, although several properties, including the processes of light radiation from the laboratory plasmoids remain to be explained.

Chapter 9

Present Aspects of the Ball Lightning Problem

Despite reports of upwards of one thousand observations in the literature and more than a half dozen comprehensive, detailed reviews of the problem, including two monograph volumes, published in the last 125 years, ball lightning remains one of the greatest mysteries of thunderstorm activity. Still less can it be said that experiment has succeeded where theory fails in duplicating more than the simplest appearance of the glowing spheres. A great number of theories have been offered, themselves involving a bewildering variety of phenomena in widely separated areas of science.

The properties which in combination present such difficulty have been noted repeatedly in observations of the fireball and are characteristic of this phenomenon alone. These spheres occur during a thunderstorm, usually associated in some way, although not necessarily directly in time or in location, with ordinary lightning. They appear as approximately globular objects 25 cm in diameter with marked colors, red, yellow, or occasionally a dazzling white, moving in the air over an extended and sometimes complex path near the ground, entering buildings, crackling noisily like an electric discharge often for a period of five seconds, and then disappearing suddenly with a loud explosion or without a sound. Reports of these properties, so different from those of the common linear flash of lightning, have met with much skepticism; but a great many detailed observations are recorded, and some photographs have been obtained.

Continuing investigation of ball lightning has succeeded in providing reasonable explanations for some of the characteristics associated with the natural phenomenon. The properties displayed by these glowing spheres vary widely in numerous occurrences, and the explanations proffered, while seemingly effective and appropriate for specific cases, also show a wide dissimilarity. The diversity in appearance and behavior of ball lightning in different cases has led to the conclusion that different types of ball lightning may exist. This view is evidently supported by the results of the considerations presented here and their correlation with the natural object.

It is thus conceivable that no single theory will account for all the natural occurrences of fireballs in a complete and conclusive way. On the other hand, certain general characteristics are displayed in many of the reported cases. The general concept of ball lightning as a plasmoid, if extended to include somewhat lower temperature and degree of ionization than usual, provides a useful description which by different charge density, temperature, composition, and structure can account for the many differing characteristics observed. This description is often appropriate regardless of the means by which the ball is generated in specific cases.

A few processes for the formation of ball lightning in nature find substantial support from direct observations and from experimental studies in the laboratory. The formation and detachment of individual luminous spherical masses from linear lightning flashes has been observed directly in thunderstorms more than once. There is no evident association of ball lightning and bead lightning in this process, for the occurrence of one seems to exclude the other. The relatively rare appearance of mobile, soap-bubble-like balls of faint luminosity following lightning activity near marshes or natural vegetation may be accounted for by the diffuse combustion-oxidation of gases ignited by an ordinary flash. Glowing balls of this type have been produced in the laboratory. The vaporization of metal forming a fireball as the result of a flash of ordinary lightning or a strong electrical discharge in the laboratory has been observed in several instances. The characteristics of the resulting spheres are in good agreement with the properties of ball lightning most difficult to explain, such as the motion and long existence. On the other hand, the metal-vapor objects are themselves not well understood. The density of charged particles and the structure of such globes have not been established. It should be noted that the luminosity of the bright spheres in general may result from excited electronic levels not requiring ionization of the material in the globe. The long life and moderate structural stability of the metal-vapor balls formed in laboratory experiments show that the data on loss of energy from such masses and on their internal structure is incomplete. The standing-wave theories, in which ball lightning is ascribed to a high-frequency external radio wave, remain of great interest. Additional information on the production of plasmoids under conditions approaching those which would exist in nature has been obtained since this process was first considered. Relatively modest power (of the order of 500 W) well within the quantities liberated by the electrical activity of storms, required, however, in a single frequency, has been found sufficient to maintain a plasmoid in air. The presence in nature of the high-frequency standing waves with the field strength required by the theory has not been established. Experimental study of this theory has been particularly limited, with little consideration of the basic parameters presented in the model.

The basic processes involved in ball lightning according to these theories are evidently readily subject to laboratory investigation. Despite numerous studies in which luminous spheres were formed by the general methods described, only a limited number of the specific characteristics associated with ball lightning have been reproduced, for example, the general appearance. Detailed measurements of the physical and chemical properties of luminous balls generated according to the methods described would contribute significantly in consideration of the theories of ball lightning. Experimental reproduction of ball lightning can be tested by simple criteria among which are the generation of the glowing sphere in air at normal pressure, existence of the fireball for several seconds at an appreciable distance from its origin or from the source of energy, and motion of the ball over an extended path.

References

The citation of different journals under a single reference number indicates republication of the same material, sometimes in the form of an abstract. The reader may select the most readily accessible journal or convenient language. A reference with papers lettered in sequence a, b, c, etc., indicates progressive publications by a given author on a subject with expansion or additional data in the subsequent work. The lettered sequence is also used for series of papers, for example, the review by de Jans which appeared in several installments. The publications by a given (first) author are referenced chronologically regardless of coauthors.

1. A. d'Abbadie, *Compt. rend.* **34,** 894 (1852).
2. C. G. Abbot, *Smithsonian Ins. Misc. Collections* **92,** No. 12 (1934).
3. D. F. Adamson, *Electrician* **25,** 445 (1890).
4. P. d'Alcantara, *Compt. rend.* **109,** 496 (1889).
5. P. d'Alcantara, *Compt. rend.* **111,** 496 (1890).
6. G. Aliverti and G. Lovera, *Arch. Meteorol. Geophys. Bioklimatol. A* **3,** 77 (1950).
7. E. Alt, *Meteorol. Z.* **18** 573 (1901).
7a. M. D. Altschuler, L. L. House, and E. Hildner, *Nature* **228,** 545 (1970).
8. W. H. Andersen, *J. Geophys. Res.* **70,** 1291 (1965).
9. W. H. Andersen, *J. Geophys. Res.* **71,** 680 (1966).
10. O. Anderson, W. R. Baker, A. Bratenahl, H. P. Furth, and W. B. Kunkel, *J. Appl. Phys.* **30,** 188 (1959).
11. R. Anderson, S. Bjornsson, D. C. Blanchard, S. Gathman, J. Hughes, S. Jónasson, C. B. Moore, H. J. Survilas, and B. Vonnegut, *Science* **148,** 1179 (1965).
12. *Anglo-Saxon Chronicle,* Jan, 793 A. D.; cf. J. Algeo and T. Pyles, *Problems in the origins and Development of the English Language,* Harcourt-Brace and World, Inc., New York, 1966. p. 126.
13. C. A. Angstrom, *Öfversigt Kongl. Vetenskaps-Akadem. Förhandl. (Stockholm)* **40,** fasc. 7, 87 (1883).
14. R. Aniol, *Meteorol. Rundschau* **7,** No. 11–12, 220 (1954).
15. V. I. Arabadji, *Sci. Notes Gorky Minsk Pedagog. Inst.* No. 5 (1956).
16. F. Arago, a) *Annuaire,* Bureau des Longitudes, 1838; b) *Oeuvres de Francois Arago,* Claye, Paris, 1854. Vol 4, p. 37 *Sämtliche Werke,* Weibel, Leipzig, 1854. Vol. 4, p. 45; *Meteorological Essays,* Longman, Brown, Green, and Longmans, London, 1855.
17. Aristotle, *Meteorologica,* trans. E. W. Webster, Oxford Univ. Press, Oxford, 1923, Book III, 1.
18. S. A. Arrhenius, *Lehrbuch der kosmischen Physik I,* S. Hirzel, Leipzig, 1903. p. 772.
19. S. E. Ashmore, *Quart. J. Roy. Meteorol. Soc.* **66,** 194 (1940).

20. G.A. Askaryan et al, *Zh. Eksperim. Teor. Fiz.–P.R.* **5**, 150 (1967); *J. Exp. Theor. Phys. Letters.* **5**, 121 (1967).
21. *L'Astronomie* **5**, 432 (1886).
22. *L'Astronomie* **9**, 312 (1890).
23. *L'Astronomie* **10**, 77 (1891).
24. *L'Astronomie* **10**, 357 (1894).
25. *L'Astronomie* **13**, 319 (1894).
26. M. Audoin, *Bull. Soc. Astron. France (L'Astronomie)* **26**, 436 (1912); *Meteorol. Z.* **30**, 148 (1913).
27. G. I. Babat, *J. Inst. Elect. Eng.* **94**, Part III, 27 (1947).
28. Babick, *Z. Meteor.* **9**, 378 (1955).
29. Babinet, *Compt. rend.* **35**, 1 (1852).
30. A. Bagate, *Bull. Soc. Astron. France* **33**, 284 (1919).
31. R. M. L. Baker, *J. Astronaut.* **15**, 44 (1968).
32. V. V. Balyberdin, *Samolet. i tekh. vozdush. flota (Airplanes and techn. air fleets)* No.3, 102 (1965).
33. V. V. Balyberdin, *Samolet. i tekh. vozdush. flota* No. 5, 3 (1966).
34. M. Baratoux, *La Météorologie* **1**, No. 28, 164 (1952).
35. J. D. Barry, a) *Wiss. Z. Elektrotech. Hochsch. Ilmenau* **9**, 202 (1967); b) *J. Atmos. Terres. Phys.* **30**, 313 (1968).
36. J. D. Barry, *J. Atmos. Terres. Phys.* **29**, 1095 (1967).
37. C. Bauer, *Umschau* **42**, 710 (1938).
38. G. Baumann, *Meteorol. Z.* **54**, 192 (1937).
39. I. Bay, *Compt. rend.* **146**, 554 (1908); *Bull. Soc. Astron. France* **22**, 232 (1908); *Meteorol. Z.* **25**, 468 (1908).
40. D. G. Beadle, *Nature* **137**, 112 (1936).
41. Du Bellay, *L'Astronomie* **5**, 311 (1886).
42. C. Benedicks, *Arkiv. Geofysik.* **2**, 1 (1951).
43. K. Berger, in F. W. Lane, *The Elements Rage,* Chilton, New York, 1965. p. 134.
44. K. Berger, *J. Franklin Inst.* **283**, 478 (1967).
45. H. P. Berlage, *Hemel en Dampkring* **53**, 65 (1955).
46. P. Bertholon, *De l'Electricité des Météores,* Croullebois, Paris, 1787, in *Landmarks of Science., Vol. 2,* p. 27.
47. Besnou, *Mem. Soc. Sci. Nat. Cherbourg* **1**, 103 (1852).
48. E. Biji, *Ciel et Terre* **26**, 246 (1905).
49. E. Blanc, *Compt. rend.* **84**, 666 (1877); *Nature* **15**, 539 (1877).
50. F. Blumhof, *Meteorol. Z.* **22**, 132 (1905).
51. K. Boll, *Wetter (Z. angew. Meteorol),* **35**, 130 (1918).
52. K. Boll, *Wetter* **37**, 191 (1920).
53. L. C. W. Bonacina, *Weather* **1**, 122 (1946).
54. A. Bonney, *Amer. Met. J.* **4**, 148 (1887).
55. H. A. H. Boot and R. B. R. Shersby-Harvie, *Nature* **180**, 1187 (1957).
56. H. A. H. Boot, S. A. Self, and R. B. R. Shersby-Harvie, *J. Electron. Control* **4**, 434 (1958).
57. W. Borlase, *Phil. Trans. Roy. Soc. London* **48**, 86 (1753); **10**, 335 (1809 abridg.)
58. W. H. Bostick, *Phys. Rev.* **104**, 292 (1956).
59. W. H. Bostick, *Phys. Rev.* **106**, 404 (1957).
60. C. M. Botley, *Weather* **21**, 318 (1966).
61. K. F. Bottlinger, *Naturwiss.* **16**, 220 (1928).
62. Bougon, *Bull. Soc. Astron. France* **16**, 420 (1902).
63. R. Boyle, *The Philosophical Works of Robert Boyle, Vol. III,* 2nd Ed., Innys and

Manby and Longnan, London, 1738. p. 32.
64. C. M. Braams, W. J. Schrader, and J. C. Terlouw, *Nucl, Instr. Methods* **4**, 327 (1959).
65. W. Brand, *Der Kugelblitz,* Henri Grand, Hamburg, 1923.
66. A. Brash, F. Lange, and C. Urban, *Naturwiss.* **16**, 115 (1928).
67. O. S. Brereton, *Phil. Trans. Roy. Soc. London* **71**, 42(1781); **15**, 21 (1809 abridg).
68. M. Brook, G. Armstrong, R. P. H. Winder, B. Vonnegut, and C. B. Moore, *J. Geophys. Res.* **66**, 3967 (1961).
69. M. Brook and N. Kitagawa, *J. Geophys. Res.* **69**, 2431 (1964).
70. S. Broughton, *Nature* **7**, 416 (1873).
71. G. H. Brown, *Meteorol. Mag.* **86**, 375 (1957).
72. T. Browne, *The Miscellaneous Writings of Sir Thomas Browne,* Faber and Faber, London, 1946. p. 195.
73. C. E. R. Bruce, *J. Inst. Electr. Eng.* **9**, NS, 357 (1963).
74. C. E. R. Bruce, *Engineer* **216**, 1047 (1963).
75. C. E. R. Bruce, *Nature* **202**, 996 (1964).
76. Bruly-Mosle, *Bull. Soc. Astron. France* **22**, 528 (1908).
77. W. Brzak, *Meteorol. Z.* **9**, 355 (1892).
78. G. Budde, *Wetter* **37**, 87 (1920).
79. *Bull. Soc. Astron. France* **14**, 510 (1900).
80. *Bull. Soc. Astron. France* **20**, 100 (1906).
81. Butti, *Compt. rend.* **35**, 193 (1852).
82. E. Caballero, *Nature* **41**, 303 (1890); *Meteorol. Z.* **7**, 158 (1890); *Ciel et Terre* **11**, 250 (1890); *La Nature* **18**, 167 (1890).
83. L. Cabane, *L'Astronomie* **6**, 459 (1887).
84. Cadenat, *Compt, rend,* **111**, 492 (1890); *Wetter* **7**, 285 (1890).
85. V. Carlheim-Gylensköld, *Ber. Internat. Meteorol. Direktorenkonf. Innsbruck,* 1905, App. 23, 1–3; *K. Z. Meteorol. Geodyn. Anh. Jahrbuch,* 1905. p. 113–115.
86. D. G. Carpenter, *Plasma Theory Applied to Ball Lightning,* Iowa State Univ., 1962. Thesis No. 62–4145, Univ. Microfilms, Ann Arbor, Michigan.
87. D. G. Carpenter, *AIAA Student J.* **1**, 25 (1963).
88. C. Carré, *Lumière électr.* **33**, 143 (1889).
89. J. Carruthers, *Meteorol. Mag.* **76**, 210 (1947).
90. Cartwright, *J. Soc. Tel. Eng. (J. Inst. Elect. Eng.)* **1**, 372 (1872).
91. B. W. Cartwright, *Life* **3**, No. 8, 77 (1937).
92. W. Cawood and H. S. Patterson, *Nature* **128**, 150 (1931).
93. H. Cecil, *Nature* **30**, 289 (1884).
94. DeCerfz, *Compt. rend.* **25**, 85 (1847).
95. M. Cerrillo, *Comision Impulsora y Coord. de la Investig. Cient.* 151 (1945), Mexico, D. F.
96. Chalmers, *Phil. Trans. Roy. Soc.* **46**, 366 (1750); **10**, 19 (1809 abridg.).
97. J. A. Chalmers, *Atmospheric Electricity,* Pergamon, New York, 1957. p. 255; 2nd Ed., 1967. p. 390.
98. S. Chandrasekhar, *Stellar Structure,* Univ. of Chicago, Chicago, 1939; Dover Publications Inc., N.Y., 1957. pp. 84–182.
99. P. N. Chirvinskiy, *Klimat i Pogoda (Climate and Weather)* **68**, No. 5 (1936).
100. P. N. Chirvinskiy, *Meteorologiya i Gidrologiya,* **7**, 78 (1936).
101. P. N. Chirvinskiy, *Priroda* **38**, No. 6, 14 (1949).
102. P. N. Chirvinskiy, *Priroda* **43**, No. 8, 116 (1954).
103. P. Clare, *Phil. Mag.* **37**, 329 (1850).
104. G. de Claubry, *Compt. rend.* **79**, 137 (1874).
105. J. Clavel, *Ciel et Terre* **22**, 302 (1901).

106. J. D. Cobine and D. A. Wilbur, *J, Appl. Phys.* **22**, 835 (1951).
107. A. E. Cocking, *Nature* **30**, 269 (1884).
108. S. C. Coroniti. *Problems of Atmospheric and Space Electricity,* Elsevier, Amsterdam, 1965.
109. Coulvier-Gravier, *Recherches sur les Météores et sur les lois qui les regissent,* Paris, 1859, p. 185 et seq.
110. J. D. Craggs and J. M. Meek, *Proc. Roy. Soc. London* A **186**, 241 (1946).
111. J. D. Craggs, W. Hopwood, and J. M. Meek, *J. Appl. Phys.* **18**, 919 (1947).
112. H. Craigie, *Science* **72**, 344 (1930).
113. W. Crookes, *Chemical News* **65**, 301 (1892).
114. A. W. Crossley, *Nature* **114**, 10 (1924).
115. Cunisset-Carnot, *Bull. Soc. Astron. France (L'Astronomie).* **11**, 299 (1897).
116. V. Cushing and M. S. Sodha, *Phys. Fluids* **2**, 494 (1959); **3**, 142, 489 (1960).
117. P. Dalloz, *La Montagne,* No. 210, 81 (1928).
118. J. Datlov, *Czech. J. Phys.* **B15**, 858 (1965).
119. G. A. Dawson and R. C. Jones, "Ball Lightning as a Radiation Bubble," in *Planetary Electrodynamics,* S. C. Coroniti and J. Hughes, ed., Gordon and Breach, New York, 1969.
120. J. D. Daugherty and R. H. Levy, *Phys. Fluids* **10**, 155 (1967).
121. A. Dauvillier, *Compt. rend.* **245**, 2155 (1957).
122. A. Dauvillier, *Compt. rend.* **260**, 1707 (1965).
123. B. Davidov, *Priroda* **47**, No. 1, 96 (1958).
124. C. Decharme, *Compt. rend.* **98**, 606 (1884); *Lumière élect.* **11**, 551 (1884).
125. H. Dember and U. Meyer, *Meteorol. Z.* **29**, 384 (1912).
126. J. Dessens, *J. Rech Atmos.* **2**, 91 (1965).
127. F. E. Dixon, *Weather* **10**, 98 (1955).
128. M. T. Dmitriev, a) *Priroda* **56**, No. 6, 98 (1967); b) *Zh. Tekhn. Fiz.* **39**, 387 (1969); *Soviet Phys.—Tech. Phys.* **14**, 284 (1969).
129. V. Dobelmann, *L'Astronomie* **25**, 262 (1911).
130. A. E. Dolbear, *Science* **11**, 38 (1888).
131. H. Dolezalek, *Geofisica pura e applicata* **20**, 183 (1951).
132. G. G. Dolgov-Savel'ev, V.S. Mukhovatov, V. S. Strelkov, M. N. Shepelev, and N. A. Yavlinskii, *J. Exp. Theor. Phys. (USSR)* **38**, 394 (1960); *Soviet Phys.—JETP* **11**, 287 (1960).
133. R. A. Dory, D. W. Kerst, D. M. Meade, W. E. Wilson, and C. W. Erickson, *Phys. Fluids* **9**, 997 (1966).
134. H. Ducros, *Bull. Soc. Astron. France* **18**, 476 (1904).
135. Dunn, *Les Nouveautés photographiques,* Libraire Illustré Paris, 1894. pp. 253, 259.
136. H. C. Early and W. G. Dow, *Phys. Rev.* **79**, 186 (1950).
137. H. S. Eaton, *Amer. Met. J.* **4**, 148 (1887).
138. Edlund, *Öfversigt Kongl. Vetenskaps-Akadem. Förhandl. (Stockholm)* **40**, fasc 7, 86 (1883).
139. H. N. Ekvall, *Electrical World* **147**, 85 (1957).
140. Emde, *Wetter* **6**, 68 (1889).
141. M. Ericson, C. S. Ward, S. C. Brown, and S. J. Buchsbaum, *J. Appl. Phys.* **33**, 2429 (1962).
142. Mme Espert, *Compt. rend.* **35**, 192 (1852).
143. J. P. Espy, *Philosophy of Storms,* Little and Brown, Boston, 1841. p. 266.
144. W. H. Evans and R. L. Walker, *J. Geophys. Res.* **68**, 4455 (1963).
145. Ezekiel, Bible, Ch. I. verses 1–28.
146. M. F. Falkner, *Meteorol. Mag.* **93**, 95 (1964).

References

147. Michael Faraday, *Experimental Researches in Electricity,* Vol, 1, Bernard Quaritch, London, 1839. p. 523.
148. H. Faye, *Compt. rend.* **111,** 492 (1890); *Wetter (Z. angew. Meteorol.)* **7,** 285 (1890).
149. H. Faye, *L'Astronomie* **10,** 22 (1891).
150. U. Fehr, *Fire Ball,* M. S. Thesis, Hebrew University, Jerusalem, 1962.
151. D. Finkelstein and J. Rubenstein, *Phys. Rev.* **135,** A390 (1964).
152. M. Fitzgerald, *Quart. J. Meteorol. Soc.* **4,** 160 (1878).
153. C. Flammarion, *The Atmosphere,* Harper and Brothers, New York, 1873. Chapt. 6; *L'Atmosphere,* Libraire Hachette, Paris, 1872, 1888.
154. C. Flammarion, *Bull. Soc. Astron. France* **13,** 145 (1899).
155. C. Flammarion and M. Fouche, *Bull. Soc. Astron. France* **18,** 378 (1904).
156. H. T. Flint, *Quart. J. Roy. Meteorol. Soc.* **65,** 532 (1939).
157. W. de Fonvielle, *Eclairs et Tonnerre,* Hachette, Paris, 1867; *Thunder and Lightning,* Sampson, Low, Son, and Marston, London 1868. pp. 32–39.
158. T. Forster, *Research about Atmospheric Phaenomena,* 3rd ed., Harding, Mavor, and Lepard, London, 1823. pp. 214–216.
159. B. D. P. Foster, *Meteorol. Mag.* **76,** 210 (1947).
160. G. Fougères, *Fulmen,* in *Dictionnaire des Antiquités grecques et romaines,* C. Daremberg and E. Saglio, ed., Librairie Hachette, Paris, 1896. Vol.2, Part II, p. 1352.
161. G. D. Freier, *J. Geophys. Res.* **65,** 3504 (1960).
162. Y. I. Frenkel, *J. Exp. Theor. Phys. (USSR)* **10,** 1424 (1940).
163. Galle, *Wetter* **7,** 238 (1890).
164. I. Galli, *Mem. Pontif. Accad. Rom. Nuovi Lincei,* a) **26,** 43 (1908); b) **28,** 217 (1909); c) **28,** 55 (1910); d) **29,** 251 (1910); e) **29,** 367 (1911); f) **30,** 257 (1911); g) **31,** 225 (1912).
165. I. Galli, *Atti Pontif. Accad. Rom. Nuovi Lincei* a) **62,** ϕ. VIII, 2 Mar. (1909); b) **62,** 18 Apr. (1909); c) **63,** 3rd note, p. 11, 17 Apr. (1910); d) **63,** 12 June (1910); e) **64,** 15 Jan (1911); f) **65,** 6th note, p. 3 (1912).
166. I. Galli, *Boll. Soc. sismolog. Ital.* **14.** 221 (1910).
167. J. W. Gallop, T. L. Dutt, and H. Gibson, *Nature* **188,** 397 (1960).
168. R. Garreau, *Compt. rend.* **209,** 60 (1939).
169. H. Gathemann, *Ann. Hydrog.* **17,** 39 (1889); *Wetter* **6,** 90 (1889).
170. J. Geerk and H. Kleinwachter, *Z. Physik.* **159,** 378 (1960).
171. F. Geist, *Umschau* **49,** 255 (1949).
172. R. Geller and M. Lucarain, *Compt. rend.* **253,** 1542 (1961).
173. T. F. Gerasimenko, *Priroda,* No. 7, 109 (1956).
174. E. Gerjuoy and R. C. stabler, *Phys. Fluids* **7,** 920 (1964).
175. W. Gerlach, *Naturwiss.* **15,** 522 (1927).
176. O. Gilbert, *Die meteorologischen Theorien des griechischen Alteraums,* B. G. Teubner, Leipzig, 1907. pp. 635–636.
177. G. Gilmore, *Nature* **103,** 284 (1919).
178. P. Göbel, *Wetter* **6,** 68 (1889).
179. P. Göbel, *Wetter* **6,** 237 (1889).
180. A. Gockel, *Meteorol. Z.* **26,** 458 (1909).
181. A. Gockel, *Das Gewitter,* F. Dummlers Verl., Berlin, 1925.
182. W. H. Godwin-Austin, *Nature* **28,** 173 (1883).
183. E. Gold, *Nature* **169,** 561 (1952).
184. B. L. Goodlet, *J. Inst. Elect. Eng.* **81,** 1 (1937).
185. H. Grad, *Phys. Fluids* **10,** 137 (1967).
186. Grobe, *Meteorol. Z.* **44,** 312 (1927).
187. V. A. Gubichev, *Ulyanovskiy gosud. pedag. inst.,* Ser. Tekh, **20,** 203 (1966); *Foreign Sci. Bull. (Library of Congress)* **4,** No. 6, 27 (1968).

188. A. Guillemin, *Le Monde Physique* **3**, 429 (1883).
189. R. Gunn, *Science* **150**, 888 (1965).
190. Haag, *Wetter(Z. angew. Meteorol.)* **31**, 240 (1914).
191. C. M. Haaland, *Phys. Rev. Letters* **4**, 111 (1960).
192. W. von Haidinger, *Sitzber. K. Akad. Wiss., Wien, Math.-Naturwiss. Kl.* **58**, (II), 761 (1868).
193. W. von Haidinger, *Sitzber. K. Akad. Wiss., Wien, Math.-Naturwiss. Kl.* **58** (II). 1045 (1868).
194. Halluitte, *Bull. Soc. Astron. France* **23**, 461 (1909).
195. C. W. Hamitlton, *Nature* **188**, 1098 (1960).
196. Sir W. Hamilton, *Phil. Trans. Roy. Soc. London* **85**, 73 (1795).
197. A. Hands, *English Mech.* **90**, 40 (1909); *Quart. J. Roy. Meteorol. Soc.* **35**, 301 (1909); *Ciel et Terre* **31**, 383 (1910).
198. C. Hare, *Nature* **19**, 5 (1878).
199. A. T. Hare, *Nature* **40**, 415 (1889).
200. Sir W. Snow Harris, *On the Nature of Thunderstorms,* J. W. Parker, London, 1843,
201. E. R. Harrison, *J. Electron. Control* **5**, 319 (1958).
202. A. C. Hawkins, *Bull. Am. Meteorol. Soc.* **25**, 364 (1944); **27**, 5 (1946).
203. E. Hayden, *Science* **10**, 324 (1887).
204. E. Hayden, *Science* **11**, 110 (1888).
205. K. G. Hernqvist, *J. Appl. Phys.* **27**, 1226 (1956).
206. E. Herrich, *Kosmos* **50**, 265 (1954).
207. N. A. Hesehus, *Beibl. Ann. Phys.* **1**, 449 (1877); *Zh. Russ. Khim. Fiz. Obshch.* **8**. 311, 356 (1876).
208. N. A. Hesehus, *Nautschnoje Obosrenije* No. 7, 1373 (1989); *Beibl. Ann. Phys.* **24**, 200 (1900); *Meteorol. Z.* **17**, 382 (1900); *Phys. Z.* **2**, 578 (1901).
209. H. Hildebrand Hildebrandsson, *Ann. Soc. Météor. France (Meteorologie)* **31**, 364 (1883); *Meteorol. Z.* **2**, 118(1885).
210. H. Hildebrand Hildebrandsson, *La Nature (Paris)* **12**, 302 (1884).
211. H. Hildebrand Hildebrandsson, *Meteorol. Z.* **13**, 475 (1896).
212. E. Hill, *Nature* **56**, 293 (1897).
213. E. L. Hill, *J. Geophys. Res.* **65**, 1947 (1960).
214. R. D. Hill, *J. Geophys. Res.* **68**, 1365 (1963).
215. A. Hofmann, *Z. angew. Meteorol.* **36**, 126 (1919).
216. W. Höhn, *Wetter u. Leben* **18**, 56 (1966).
217. A. Höhr, *Meteorol. Z.* **20**, 570 (1903).
218. M. Holmes, *Nature* **133**, 179 (1934).
219. R. E. Holzer and E. J. Workman, *J. Appl. Phys.* **10**, 659 (1939).
220. H. Homma, *Meteorol. Z.* **18**, 576 (1901).
221. F. Horner and P. A. Bradley, *J. Atmos. Terrest. Phys.* **26**, 1155 (1964).
222. D. W. Horner, *Symons's Meteorol. Mag.* **39**, 111 (1904).
223. W. J. Humphreys, *Physics of the Air*, McGraw-Hill, New York, 2nd ed., 1929; 3rd ed., 1940. Ch. 17, 18.
224. W. J. Humphreys, *Proc. Am. Phil. Soc.* **76**, 613 (1936).
225. M. E. Hunneman, *Science* **86**, 244 (1937).
226. Th. Indrikson, *Zh. Russ. Khim. Fiz. Obshch.* **32**, 53 (1900); *Beibl. Ann. Phys.* **24**, 706 (1900).
227. H. Israel, *Umschau* **58**, 586 (1958).
228. A. Jaeger, *Meteorol. Rundschau* **14**, 31 (1961).
229. E. Janezic, *Meteorol. Z.* **25**, 42 (1908).
230. C. de Jans, *Ciel et Terre,* a) **31**, 499 (1910); b) **32**, 155 (1911); c) **32**, 255 (1911); d) **32**,

301 (1911); e) **33**, 18 (1912); f) **33**, 143 (1912).
231. R. C. Jennings, *New Scientist* **13**, 156 (1962).
232. R. C. Jennison, *Nature* **224**, 895 (1969).
233. J. C. Jensen, *Physics* **4**, 372 (1933); *Sci .Mo.* **37**, 190 (1933); *Nature* **133**, 95 (1934).
234. E. O. Johnson, *R.C.A. Rev.* **16**, 498 (1955).
235. L. H. Johnson. *Marine Obs.* **28**, 11 (1958).
236. P. O. Johnson. *Am. J. Phys.* **33**, 119 (1965).
237. A. T. Jones, *Science (New Series)* **31**, 144 (1910).
238. N. Jones, *Nature* **130**, 545 (1932).
239. J. Kaiser, *Science* **60**, 293 (1924).
240. L. F. Kämtz, *Lehrbuch der Meteorologie,* Halle, 1832. Vol. 2, p. 427.
241. P.L. Kapitsa, *Dokl. Akad. Nauk. SSSR* **101**, 245 (1955); *Phys. Blätter* **14**, 11 (1958); *Ball Lightning*, Consultants Bureau, New York, 1961. p. 11–16
242. P. L. Kapitsa, *Uspekhi fiz. nauk* **78**, 181 (1962); *Soviet Phys.—Uspekhi* **5**, 777 (1963).
243. P. L. Kapitsa, *Zh. tekhn. fiz.* **38**, 1829 (1968); *Soviet Phys.—Tech. Phys.* **13**, 1475 (1969)
244. R. Käubler, *Phys. Blätter* **19**, 374 (1963).
245. T. Kerkhoff, *Wetter (Z. angew. Meteorol.)* **30**, 286 (1913).
246. S. Khalatow and D. Khalatow, *Bull. Soc. Astron. France* **21**, 452 (1907).
247. M. A. R. Khan, *Nature* **155**, 53 (1945).
248. E. von Kilinski, *Lehrbuch der Luftelektrizität,* Akad. Verlag & Geest and Portig K.-G., Leipzig, 1958. p. 30.
249. N. G. Kim, *J. Appl. Phys.* **36**, 1611 (1965).
250. A. Kimpara, *Electromagnetic Radiation from Lightning,* in S. C. Coroniti, *Problems of Atmospheric and Space Electricity,* Elsevier, New York, 1965. pp. 352-364.
251. E. King, *Phil. Trans. Roy. Soc. London* **63**, 231 (1773); **13**, 435 (1809 abridg.).
252. H. R. Kingston, *J. Roy. Astron. Soc. Canada* **25**, 420 (1931).
253. N. Kitagawa, M. Brook, and E. J. Workman, *J. Geophys. Res.* **67**, 637 (1962).
254. J. Klose, *Z. Meteorol.* **4**, 91 (1950).
255. W. Knoche, *Meteorol. Z.* **26**, 355. (1909).
256. J. Koechlin, *La Nature (Paris)* **52**, 12 July, No. 2628 (1924).
257. G. I. Kogan-Beletskii, *Priroda* **46**, No. 6, 71 (1957); transl. in *Ball Lightning,* D, J. Ritchie, ed., Consultants Bureau, New York, 1961. pp. 29–32.
258. T. Köhl, *Gaea* **18**, 569 (1882).
259. T. Köhl, *Klein's Wochenschr, Astron., Meteorol. Geog.* **27**, 140 (1884).
260. N. V. Kolobkov, *Grozy i Shkvaly (Thunderstorms and Squalls)* Gos. Izdat. Tekh.- Teoret. Lit., Moscow, 1951. pp. 47–50
261. E. L. Kosarev et al., *Zh. tekhn. fiz.* **38**, 1831 (1968); *Soviet Phys.—Tech. Phys.* **13**, 1477 (1969).
262. M. Kruskal and M. Schwarzschild, *Proc. Roy. Soc. London* **A223**, 348 (1954).
263. E. Kuhn, *Naturwiss.* **38**, 518 (1951).
264. W. B. Kunkel, *J. Appl. Phys.* **21**, 820 (1950).
265. Kuschewitz, *Elektrotech. Z.* **26**, 829 (1905).
266. Yu. P. Ladikov, *Izvest. Akad. Nauk SSSR, Otdel. Tekhn. Nauk. Mekh. i Mashin.* No. 4, 7 (1960); transl. in *Ball Lightning,* D. J. Ritchie, ed., Consultants Bureau, New York, 1961. pp. 51–59.
267. E. Lagrange, *Ciel et Terre* **30**, 417 (1909).
268. E. Lagrange, *Ciel et Terre* **31**, 385 (1910.).
269. De Lalande, *Compt. rend.* **35**, 24 (1852); **35**, 400 (1852).
270. A. Lalung. *Bull. Soc. Astron. France* **13**, 500 (1899).
271. A. Lancaster, *Ciel et Terre* **26**, 246 (1905).

272. F. W. Lane, *The Elements Rage*, Chilton, New York, 1965. pp. 134–141
273. S. Leduc, *Compt. rend.* **129**, 37 (1899).
274. C. Lehideux, *Bull. Soc. Astron. France* **18**, 477 (1904).
275. B. Lehnert, *Phys. Fluids* **4**, 847 (1961).
276. B. Lehnart, *Dynamics of Charged Particles,* John Wiley and Sons, Inc., New York, 1964.
277. R. A. Leonov, *The Enigma of Ball Lightning,* Izd. "Nauka," Moscow, 1965. Transl. TT: 66-33253, Clearinghouse Fed. Sci. Tech. Inform., US. Dept. Commerce, Springfield, Virginia.
278. F. von Lepel, *Meteorol. Z.* **7**, 297 (1890).
279. E. Less, *Meteorol. Z.* **18**, 39. (1901).
280. B. Lewis and G. von Elbe, *Combustion, Flames, and Explosions of Gases,* Academic Press, New York, 1961.
281. H. D. Lewis, *Sci. Amer.* **208**, 107 (1963).
282. *Ball Lightning Bibliography:* 1950–1960, Library of Congress, Science and Technology Div., Washington, D. C., 1961.
283. L. Lindberg, E. Witalis and C. T. Jacobsen, *Nature* **185**, 452 (1960).
284. G. Lindemann, *Kosmos* **47**, 380 (1951).
285. J. G. Linhart, *Plasma Physics,* Interscience, New York, 1960.
286. K. List, *Meteorol. Z.* **22**, 139 (1905).
287. F. G. Lloyd, *English Mech.* **45**, 159 (1887).
288. O. J. Lodge, *Lightning Conductors and Lightning Guards,* Whittaker and Co. and Bell and Sons, London, 1892. pp. 138–139.
289. L. B. Loeb, *Static Electrification,* Springer-Verlag, Berlin–Göttingen, 1958.
290. L. B. Loeb, *J. Geophys. Res.* **69**, 587 (1964).
291. L. B. Loeb, *J. Geophys. Res.* **71**, 676 (1966).
292. E. A. Logan, *Electr. Rev.* **138**, 381 (1946).
293. J. J. Lowke, M. A. Uman, and R. W. Liebermann, *Toward a New Theory of Ball Lightning,* Am. Geophys. Union Meeting, Washington, D. C., April 1969.
294. Lucretius, *De Rerum Natura,* Book VI, lines 145–378.
295. D. J. Malan, *Ann. Geophys.* **17**, 388 (1961).
296. D. J. Malan, *Physics of Lightning,* English Univ. Press, London, 1963. p. 7.
297. A. B. Mallinson, *J. Inst. Elect. Eng.* **81**, 46 (1937).
298. W. Malsch, *Meteorol. Rundschau* **9**, 150 (1956).
299. W. Malsch, *Meteorol. Rundschau* **9**, 188 (1956).
300. L. Malter, E.O. Johnson, and W. M. Webster, *R. C. A. Rev.* **12**, 415 (1951).
301. E. Mancini, *Atti Reale Accad. Lincei, Series V. Rendiconti sci. fis.* **1**. 2nd sem., 308 (1892); *Naturwiss. Rundschau* **8**, 62 (1893); *Meteorol. Z.* **10**, 157 (1893).
302. P. Z. von Manteuffel, *Umschau* **42**, 587 (1938).
303. E. W. Marchant, *Nature* **125**, 128 (1930).
304. B. Margs, *Meteorol. Rundschau* **9**, 72 (1956).
305. O. C. Marsh, *Nature* **53**, 152 (1895); *Amer. J. Sci., Ser.* 4, **1**, 13 (1896); *Meteorol. Z.* **13**, 189 (1896).
306. E. Mascart, *Compt. rend.* **111**, 496 (1890).
307. B. J. Mason, *Clouds, Rain, and Rainmaking,* Cambridge Univ. Press, Cambridge, England, 1962. Ch. 7. pp. 112–136.
308. E. Mathias, *Ann. de Phys.* 9th Ser., **5**, 365 (1916); *Monthly Weather Rev.* **44**, 516 (1916); *Compt rend.* **162**, 642 (1916).
309. E. Mathias, a) *Traité d'Électricité Atmosphérique et Tellurique,* Univ. de France, Paris, 1924. pp. 296–299; b) *Compt. rend.* **179**, 136 (1924).
310. E. Mathias, *Compt. rend.* a) **182**, 32 (1926); b) **189**, 512, 607, 813 (1929); c) **196**,

654 (1933).
311. E. Mathias, *Compt. rend.* a) **186,** 1334 (1928); b) **189,** 1049 (1929); c) **194,** 413 (1932); d) **194,** 2257 (1933); e) **197,** 962 (1933); f) **199,** 505 (1934);); g) **199,** 1083 (1934).
312. H. Mathouillot, *Compt. rend.* **146,** 603 (1908).
313. B. T. Matthias and S. J. Buchsbaum, *Nature* **194,** 327 (1962).
314. A. Maudemain, *Bull. Soc. Astron. France* **33,** 284 (1919).
315. Ch. Maurain, *La Foudre,* Armand Colin, Paris, 1948. pp. 100–104.
316. Mavrogordato, *L'Astronomie* **5,** 309 (1886).
317. Mavrogordato, *L'Astronomie* **5,** 310 (1886).
318. J. A. McClelland, *Nature* **103,** 284 (1919).
319. T. McDonough, *Science* **11,** 62 (1888).
320. W. G. McMillan, *Nature* **40,** 295 (1889).
321. J. R. McNally, Jr., Paper J-14, Amer. Phys. Soc., Div. Plasma Physics Meeting, Gatlinburg, Tenn., Nov. 2–5, 1960; *Bull. Am. Phys. Soc.* **6,** 202, Abstr. J-14 (1961).
322. A. Meissner, *Meteorol. Z.* **47,** 17 (1930).
323. T. C. Mendenhall, *Am. Meteorol. J.* **6,** 437 (1890).
324. O. Merhaut, *Naturwiss.* **32,** 212 (1944).
325. *Meteorol. Z.* **44,** 391 (1927).
326. Mettetal, *Compt. rend.* **121,** 596 (1895); *La Nature (Paris)* **23,** 2nd sem., 407 (1895).
327. A. Meunier, *Compt. rend.* **35,** 195 (1852).
328. S. Meunier, *Compt. rend.* **103,** 837 (1886); *L'Astronomie* **6,** 104 (1887); *Meteorol. Z.* **5,** 160 (1888).
329. Meusberger, *Meteorol. Z.* **21,** 283 (1904).
330. A. Mey, *Meteorol. Z.* **44,** 188 (1927)
331. A. H. Miller, C. E. Shelden, and W. R. Atkinson, *Phys. Fluids* **8,** 1921 (1965).
332. L. M. Milne-Thomson, *Theoretical Hydrodynamics, 2nd ed.,* MacMillan, New York, 1950.
333. G. M. Minchin, *Nature* **53,** 5 (1895).
334. R. V. Mitin and K. K. Pryadkin, *Zh. tekhn. fiz.* **35,** 1205 (1965); *Soviet Phys.–Tech. Phys.* **10,** 933 (1966).
335. H. Mohn, *Meteorol. Z.* **25,** 314 (1908).
336. F. B. Mohr, *Science* **151,** 634 (1966).
337. F. Moigno, *Cosmos* **14,** 672 (1859).
338. F. Möller, *Umschau* **49,** 458 (1949).
339. F. Möller, *Kosmos* **47,** 86 (1951).
340. Th. du Moncel, *Mem. Soc. Sci. Nat. Cherbourg* **2,** 85 (1854); *Compt. rend.* **38,** 408 (1854); *Meteorologie (Ann. Soc. Meteorol. France)* **2,** 57 (1854).
341. Th. du Moncel, *Notice historque et theorique sur le Tonnerre et les Eclairs,* Hachette, Paris, 1857. pp. 52, 54.
342. F. D. Moon, *Marine Obs.* **2,** 129 (1925).
343. C. B. Moore, B. Vonnegut, J. A. Machado, and H. J. Survilas, *J. Geophys. Res.* **67,** 207 (1962).
344. J. G. Moore, *J. Geophys. Res.* **68,** 1335 (1963).
345. W. Morris, *Daily Mail,* 5 Nov. 1936 (letter to the Editor).
346. F. T. Mott, *Nature* **22,** 193 (1880).
347. Ch. Mousette, *Compt. rend.* **103,** 30 (1886).
348. Ch. Mousette, *Compt. rend.* **107,** 435 (1888).
349. Müller, *La Nature (Paris)* **16,** 415 (1888).
350. D. Müller-Hillebrand, *Elektrie* **17,** 211 (1963).
351. D. Müller-Hillebrand, in *Problems of Atmospheric and Space Electricity,* S. C. Coroniti ed., Elsevier, Amsterdam, 1965. p. 457–459.

352. P. Van Muschenbroek, *Cours de Physique expérimentale et mathématique,* transl. Sigaud de la Fond, Leyden, 1769. Vol. 3, p. 406.
353. E. Nasse, *Compt. rend.* **74,** 1384 (1872).
354. H. Nauer, a) *Z. angew. Phys.* **5,** 441 (1953); b) *Umschau Fortschr. Wiss. Tech.* **56,** 75 (1956).
355. F. Neesen, *Himmel u. Erde,* **13,** 145 (1901).
356. T. Neugebauer, *Z. Physik* **106,** 474 (1937).
357. J. Neunteufl. *Wetter u. Leben* **3,** 49 (1951).
358. M. M. Newman, *J. Geophys. Res.* **65,** 1966 (1960).
359. Nicholson, *Phil. Trans. Roy. Soc. London* **64,** 350 (1774); **13,** 538 (1809 abridg.).
360. Nippoldt, *Meteorol. Z.* **33,** 475 (1916).
361. H. Norinder, *Kgl. Vetenskapssoc. Arsbok* 89–95 (1939).
362. H. Norinder, in *Problems of Atmospheric and Space Electricity,* S. C. Coroniti, ed., Elsevier, Amsterdam. 1965. p. 455.
363. G. Oltramare, *Compt. rend.* **88,** 1319 (1879).
364. A. Oriel, *Bull. Soc. Astron. France* **15,** 303 (1901).
365. R. E. Orville, *Science* **151,** 451 (1966).
366. R. E. Orville, *J. Atmos. Sci.* **25,** 827, 839, 852 (1968).
367. P. Oswalt, *Bull. Soc. Astron. France* **29,** 437 (1915).
368. M. Otto, *La Nature, (Paris)* **29,** (2), 361 (1901).
369. R. Owen, *Amer. Meteor. J.* **3,** 383 (1886).
370. L. Palazzo, *Meteorol. Z.* **20,** 188 (1903).
371. E. Parent, *Compt. rend.* **77,** 370 (1873).
372. P. V. H. Pascal, *Nouveau traité de chimie minérale,* Masson, Paris, 1956, Vol. 10, p. 368; Vol. 13, p. 266.
373. W. Paul, H. P. Reinhard, and U. von Zahn, *Z. Physik* **152,** 143 (1958).
374. N. B. Peake, *New Scientist* **13,** 334 (1962).
375. G. Pellissier, *La Nature (Paris)* **22,** (1), 323 (1894).
376. J. C. A. Peltier, *Am. J. Sci.* **38.** 73 (1840).
377. W. Peppler, *Wetter (Z. angew. Meteorol.)* **36,** 29 (1919).
378. E. Péroux, *Bull. Soc. Astron. France* **21,** 453 (1907).
379. P. Perrin, *Etudes sur les Éclairs,* Paris, 1873. pp. 77–84.
380. H. Petersen, *Geophysica* **5,** 49 (1952); *Weather* **9,** 73 (1954).
381. H. Petersen, *Weather* **9,** 321 (1954).
382. V. A. Petrzilka, *Cesk. Casopis Fys.* **A14,** No. 6, 542 (1964).
383. L. Graf Pfeil, *Der Ursprung der Meteoriten Gäo* **22,** 229 (1886).
384. R. Pflegel, *Z. Meteorol.* **10,** 347 (1956).
385. R. Phillips, a) *On Atmospheric Electricity,* Hardwicke, London, 1863. pp. 45–47; b) *Nature* **41,** 58 (1889).
386. N. D. Piltschikow, *Beibl. Ann. Pnys.* **24,** 691 (1900).
387. G. Planté, *Compt. rend.* **80,** 1133 (1875); **81,** 185 (1875).
388. G. Planté, *Compt. rend.* **83,** 321 (1876).
389. G. Planté, *Compt. rend.* **83,** 484 (1876).
390. G. Planté, a) *Compt. rend.* **85,** 619 (1877); b) **87,** 325 (1878); c) **99,** 273 (1884); *Ciel et Terre* **5,** 365 (1884).
391. G. Planté, *Phénomènes électriques de l'Atmosphère,* Paris, 1888; *Elektrische Erscheinungen der Atmosphäre,* G, Wallentin, Halle a. S., 1889.
392. A. Poey, *Compt. rend.* **40,** 1183 (1855).
393. C. H. Pollog, *Meteorol. Z.* **47,** 79 (1930).
394. C. B. Polychronakis, *Bull. Soc. Astron. France* **18,** 476 (1904).
395. C. B. Polychronakis, *Bull. Soc. Astron. France* **22,** 479 (1908).

References

396. C. Ponnamperuma and F. Woeller, *Nature* **203**, 272 (1964).
397. Y. A. Popov, *Priroda* **48**, No. 12, 111 (1959).
398. L. M. Potts, *Science* **31**, 144 (1910).
399. Pouillet, *Compt. rend.* **35**, 400 (1852).
400. E. de Poulpiquet, *La Nature (Paris)* **17**, sem. I. 39 (1889); *Wetter* **6**, 69 (1889); *Cosmos* **38**, 142 (1889).
401. R. M. Poulter, *Meteorol. Mag.* **70**, 289 (1936).
402. R. M. Poulter, a) *Weather* **9**, 121 (1954); b) *Weather* **9**, 321 (1954).
403. Poumier, *L'Astronomie* **5**, 348 (1886).
404. J. R. Powell, M. S. Zucker, J. F. Manwaring, and D. Finkelstein, *Bull. Am. Phys. Soc.* **12**, 751 Abstr. 2C-2 (1967).
405. J. R. Powell and D. Finkelstein, *Structure of Ball Lightning* in *Advances in Geophysics*, Vol. 13, Academic Press, New York, 1969; *Am. Scientist* **58**, 262 (1970).
406. C. E. Pratt, *Marine Obs.* **31**, 127 (1961).
407. E. Préaubert, *Ann. Soc. Météor. France* **52**, 270 (1904).
408. K. H. Prendergast, *Astrophys. J.* **123**, 498 (1956).
409. S. Price and E. M. Carlstead, *Monthly Weather Rev.* **94**, 272 (1966).
410. *Priroda* **51**, No. 5, 71 (1962).
411. O. Prochnow, *Erdball und Weltall*, Bermüller, Berlin, 1928.
412. O. Prochnow, *Physik. Z.* **31**, 335 (1930).
413. K. Prohaska, *Meteorol. Z.* **6**, 472 (1889).
414. K. Prohaska, *Meteorol. Z.* **17**, 331 (1900).
415. K. Prohaska, *Meteorol. Z.* **20**, 315 (1903).
416. A. Pühringer, in *Problems of Atmospheric and Space Electricity*, S. C. Coroniti, ed., Elsevier, Amsterdam, 1965. p. 460.
417. A. Pühringer, *Wetter u. Leben* **19**, 57 (1967).
418. O. Raab, *Bildmessg. u. Luftbildwes.* **15**, 57 (1940).
419. E. Racine, *Bull. Soc. Astron. France* **23**, 461 (1909).
420. W. D. Rayle, *Ball Lightning Characteristics*, National Aeronautics and Space Adiminstration, Washington, D.C., Tech. Note NASA TN D-3188, Jan. 1966.
421. T. B. Reed, *J. Appl. Phys.* **32**, 821 (1961).
422. W. Reid, *An Attempt to Develop the Law of Storms, etc.*, John Weale, London, 1838. pp. 402, 412.
423. E. Reimann, *Meteorol. Z.* **3**, 510 (1886).
424. E. Reimann, *Meteorol. Z.* **4**, 164 (1887).
425. E. Reimann, *Meteorol. Z.* **13**, 25 (1896).
426. E. Reimann, *Meteorol. Z.* **22**, 360 (1905).
427. F. Reiners, *Ann. Hydrograph. Marit. Meteorol.* **10**, 582 (1882).
428. W. C. Reynolds, *Nature* **112**, 903 (1923).
429. W. C. Reynolds, *Nature* **125**, 413 (1930).
430. W. C. Reynolds, *Nature* **128**, 584 (1931).
431. P. Richter, *Z. Angew. Meteorol.* **30**, 21 (1913).
432. A. Righi, *Atti R. Acc. Lincei, Rendiconti, IV.* **7**, 330 (1891).
433. L. Rihanek and J. Postranecky, in *Bourky a Ochrana Pred Bleskem*, Nakld. Cesk. Akad. Ved., Praha, 1957. pp. 90–91, 463–478.
434. D. J. Ritchie, *Ball Lightning*, Consultants Bureau, New York, 1961.
435. D. J. Ritchie, *J. Inst. Elect. Eng.* **9**, 202 (1963.).
436. A. de la Rive, *Traité d'Electricité théorique et appliquée*, Paris, 1858. Vol. 3, p. 197.
437. M. Roberts and I. Alexeff, *Bull. Am. Phys. Soc.* **12**, 26, Abstr. AJ4 (1967).
438. Roche, *Compt. rend.* **139**, 465 (1904); *Bull. Soc. Astron. France* **18**, 509 (1904).
439. M. Rodewald, *Z. Meteorol.* **8**, 27 (1954).

440. F. Rossmann, *Wetter u. Klima* **2**, 75 (1949).
441. L. Rotch, *Bull. Soc. Astron. France* **17**, 483 (1903); *Ciel et Terre* **24**, 544 (1903).
442. F. Roth, *Merteorol. Z.* **6**, 231 (1889).
443. U. Rousselot, *Bull. Soc. Astron. France* **18**, 527 (1904).
444. N. Rozzi, *Bull. Soc. Astron. France* **25**, 260 (1911).
445. M. P. Rudski, *Meteorol. Z.* **22**, 284 (1905).
446. S. C. Russell, *Symon's Meteorol. Mag.* **39**, 153 (1904); *Bull. Soc. Astron. France* **18**, 510 (1904).
447. G. M. Ryan, *Nature* **52**, 392 (1895); *Cosmos* **46**, 95 (1897); *Bull. Soc. Astron. France* **11**, 300 (1897).
448. Sacc, *Compt. rend.* **53**, 646 (1861).
449. St. Grosu, *Eletrotechnica (Rumanian)* **8**, No. 2, 57 (1960).
450. L. E. Salanave, in *Problems of Atmospheric and Space Electricity,* S. C. Cornoniti, ed., Elesevier, Amsterdam, 1965. p. 464.
451. J. Doyne Sartor, *Science* **143**. 948 (1964).
453. F. Sauter, *Uber Kugelblitze,* Beilage Programm Kgl. Real-Gymnasiums Ulm. I Teil: *Theorie der Kugelblitze,* 1890; II Teil: *Beispiele von Kugelblitzen,* 1892.
453. F. Sauter, *Meteorol. Z.* **12**, 241 (1895).
454. F. Scheminzky and F. Wolf, *Akad. Wiss. Wien, Sitzber.,* Abt. IIA **156**, 1 (1948).
455. F. von Schiödt, *Tidskr. f. Physik og Chemi* **2**, 242 (1893).
456. K. A. Schlobohm, *Meteorol. Rundschau* **14**, 93 (1961).
457. A. Schmauss, *Physik. Z.* **10**, 968 (1909).
458. A. Schmauss, *Meteorol. Z.* **35**, 184 (1918).
459. K. Schneidermann, *Feurwehr-Verbands-Z.* **44**, 1 (1934).
460. B. F. J. Schonland, a) *Advance. Sci.* **19**, 306 (1962); b) *Nature* **195**, 880 (1962).
461. B. F. J. Schonland, *The Flight of Thunderbolts,* Clarendon Press, Oxford, 1st edition, 1950, 2nd edition, 1964.
462. C. Schoute, *Meteorol. Mag.* **61**, 238 (1926).
463. Schrade, *Z. Meteorol.* **17**, 61 (1964).
464. M. Schrammen, *Ciel et Terre* **11**, 348 (1890).
465. H. Schwegler, *Naturwiss. Rundschau* **4**, 169 (1951).
466. E. K. Scott, *Nature* **112**, 760 (1923).
467. J. R. Scott, *Weather* **10**, 98 (1955).
468. R. H. Scott, *Quart, J. Meteorol. Soc.* **4**, 166 (1878).
469. Séguier, *Compt. rend.* **34**, 871 (1852).
470. R. Seigner, *Wetter u. Leben* **18**, 54 (1966).
471. F. Sestier, *De la Foudre, de ses formes et de ses effets,* Baillère et fils, Paris, 1866.
472. R. Seyboth, *Sci. Am.* **86**, 36 (1902).
473. V. D. Shafranov, *On Equilibrium Magnetohydrodynamic Configurations,* Proc. 3rd Inter. Congress on Ionization Phenomena in Gases, Venice, 11–15 June 1957. pp. 990–997.
474. V. D. Shafranov, a) *Zh. eksp. teor. Fiz.* **33**, 710 (1957); *Soviet Phys.—JETP* **6**, 545 (1958) b) *Zh eksp. teor. Fiz.* **37**, 1088 (1959); *Soviet Phys.—JETP* **10**, 775 (1960).
475. A. R. Shapiro and W. K. R. Watson, *Phys. Rev.* **131**, 495 (1963).
476. P. A. Silberg, *J. Appl. Phys.* **32**, 30 (1961).
477. P. A. Silberg, *J. Geophys. Res.* **67**, 4941 (1962).
478. P. A. Silberg, *J. Appl. Phys.* **35**, 2264 (1964).
479. P. A. Silberg, *A Review of Ball Lightning,* in *Problems of Atmospheric and Space Electricity,* S. C. Coroniti, ed., Elsevier, Amsterdam, 1965. pp. 436, 464.
480. G. C. Simpson, *Nature* **112**, 727 (1923); *Sci Am.* **130**, 242 (1924).
481. G. C. Simpson, *Nature* **113**, 677 (1924).

References

482. S. Singer, *Nature* **198**, 745 (1963); *Naturwiss. Rundschau* **16**, 450 (1963).
483. M. Skowronek, *Compt. rend.* **250**, 1808 (1960).
484. W. F. Smith, *Nature* **22**, 267 (1880).
485. W. G. Smith, *Nature* **30**, 241 (1884).
486. N. de Soubbotine, *Bull. Soc. Astron. France* **16**, 117 (1902).
487. Steinheim, *Compt. rend.* **36**, 744 (1853).
488. I. S. Stekolnikov, *Study of Lightning and Lightning Protection, ("Izucheniye Molnii i Grozozashchita")*, Izd. Akademii Nauk; Moscow, 1955: N65-20412, April, 1965, JPRS 29, 407, Clearinghouse Fed. Sci. Tech. Inform., U.S. Dept. Commerce, Springfield, Virginia.
489. I. S. Stekolnikov, *Priroda* **47**, No. 1. 96 (1958).
490. J. Studer, *Ann. Schweizer Met. Cent. Anst.* **22**, Append. 3 (1885); *Meteorol. Z.* **5**, 159 (1888); *Ann. Soc. Météorol. France* **37**, 167 (1889).
491. A. Subramanian, *Indian J. Meteorol. Geophys.* **14**, 358 (1963).
492. W. Swain, *Roy. Astron. Soc. Canada* **27**, 255 (1933).
493. C. D. Swart, *Am. Meteorol. J.* **4**, 98 (1887).
494. G. J. Symons, *Quart. J. Roy. Meteorol. Soc.* **4**, 165 (1878).
495. P. G. Tait, *Nature* **22**, 408 (1880); *Thunderstorms* in *Life and Scientific Work of Peter Guthrie Tait* by C. G. Knott, Cambridge Univ. Press, 1911. Supplement, pp. 312–314.
496. P. G. Tait, *Nature* **22**, 436 (1880).
497. P. G. Tait, *Sci. Am.* **73**, 405 (1895).
498. C. F. Talman, *Am. Mercury* **26**, 69 (1932).
499. T. L. Tanton, *J. Roy. Astron. Soc. Canada* **12**, 530 (1918).
500. De Tastes, *Ann. Soc. Météor. France (Météorologie)* **32**, 105 (1884); *Meteorol. Z.* **2**, 115 (1885).
501. J. E. Taylor, *Weather* **9**. 321 (1954).
502. M. Teich, *Z. Meteorol.* **9**, 379 (1955).
503. G. S. Teletov. *Priroda* **55**, No. 9, 84 (1966).
504. T. Terada, *Bull. Earthquake Res. Inst. Tokyo Univ.* **9**, 225 (1931).
505. J. E. TerGouw, *Bull. Soc. Astron. France* **21**, 313 (1907).
506. De Tessan, *Compt. rend.* **49**, 189 (1859).
507. J. B. Thate, *Hemel en Dampkring* **50**, 134 (1952).
508. H. Théron, *Bull. Soc. Astron. France (L'Astronomie)* **11**, 299 (1897).
509. D. Thompson, *Quart. J. Roy. Meteor. Soc.* **71**, 39 (1945).
510. E. Thomson, *Science, (New Series)* **30**, 857 (1909).
511. W. Thomson, Lord Kelvin, *Report of the 58th Mtg. British Assoc., Adv. Sci.* **58**, 604 (1888).
512. W. M. Thornton, *Phil. Mag.* **21**, 630 (1911); *Radium* **8**, 397 (1911); *Meteorol. Z.* **29**, 39 (1912); *Cosmos* **18**, sem. 1, 142 (1913).
513. G. Tissandier, *Compt. rend.* **113**, 421 (1891).
514. M. Toepler, *Ann. Phys. IV* **2**, 560 (1900); *Meteorol. Z.* **17**, 543 (1900).
515. M. Toepler, *Ann. Phys. IV* **6**, 339 (1901).
516. M. Toepler, *Meteorol. Z.* **18**, 533 (1901).
517. M. Toepler, *Z. Tech. Phys.* **10**, 73, 113 (1929).
518. M. Toepler, *Naturwiss. Rundschau* **7**, 326 (1954).
519. C. Tomlinson, *Phil. Mag.* (ser. 5) a) **26**, 114 (1888); b) **26**, 475 (1888).
520. L. Tonks, *Nature* **187**, 1013 (1960).
521. G. F. Townsend, *Elec. Rev.* **26**, 297 (1895).
522. E. Trapp, *Wetter u. Leben* **1**, 274 (1948).
523. A. Trécul, *Compt. rend.* **83**, 478 (1876).
524. A. Trécul, *Compt. rend.* **92**, 775 (1881); *L'Astronomie* **6**, 107 (1887).

525. J. R. Trimmier and A. Miller, *Phys. Fluids* **9**, 1997 (1967).
526. J. W. Tripe, *Quart. J. Met. Soc.* **2**, 431 (1875); *Amer. Met. J.* **4**, 149 (1887); *Electrician* **19**, 179 (1889).
527. J. Trowbridge, *Sci. Amer.* **96**, 489 (1907).
528. A. Turpain, J. Phys. *Théorique Appliquée* V **1**, 372 (1911).
529. M. A. Uman, *J. Atmos. Terr. Phys.* **24**, 43 (1962).
530. M. A. Uman, *J. Geophys Res.* **69**, 583 (1964).
531. M. A. Uman, R. E. Orville, and L. E. Salanave, *J. Atmos. Sci.* **21**, 306 (1964).
532. M. A. Uman and C. W. Helstrom, *J. Geophys. Res.* **71**, 1975 (1966).
533. M. A. Uman, *J. Atmos. Terrest. Phys.* **30**, 1245 (1968).
534. M. A. Uman, *Lightning,* McGraw-Hill, New York, 1969.
535. M. A. Uman, *Decaying Lightning Channels, Bead Lightning, and Ball Lightning,* in *Planetary Electrodynamics,* S. C. Coroniti and J. Hughes, eds., Gordon and Breach, New York, 1969.
536. F. C. Van Dyck, *Science* **11**, 110 (1888).
537. G. Vannesson, *L'Astronomie* **5**, 431 (1886).
538. Van Marum, *Phil. Mag.* **8**, 313 (1800).
539. C. F. Varley, *Proc. Roy. Soc. London* **19**, 236 (1871).
540. A.A. Vedenov, V. M. Glagolev, *et al., Proceedings of the Second United Nations International Conference on the Peaceful Uses of Atomic Energy,* United Nations, Geneva, 1958. Vol 32, p. 239.
541. Verdet, *La Nature (Paris)* **18**, 303 (1890).
542. P. E. Viemeister, *Lightning Book,* Doubleday and Co., Inc., New York, 1961.
543. G. G. de Villemontée, *Compt. rend.* **155**. 1567 (1912).
544. J. Violle, *Compt. rend.* **132**, 1537 (1901); *Ciel et Terre* **22**, 479 (1901); *Naturwiss. Rundschau* **16**, 504 (1901); *Meteorol. Z.* **19**, 335 (1902).
545. J. Violle, *Compt. rend.* **158**, 1542 (1914).
546. S. W. Visser, *Hemel en Dampkring* **55**, 45 (1957); *Meteorol. Mag.* **86**, 344 (1957).
547. A. A. Vlasov, a) *Scientia Sinica* **8**, 266 (1959); b) *Zh. tekhn. fiz.* **31**, 785 (1961); *Soviet Phys.—Tech. Phys.* **6**, 569 (1962.).
548. A. Voller, *Elektrotech. Z.* **9**, 473 (1888).
549. C. R. Volmer, *Wetter* **13**, 185 (1896); *Meteorol. Z.* **14**, 34 (1897).
550. B. Vonnegut and C. B. Moore, *Recent Advances in Atmospheric Electricity,* Pergamon Press, New York, 1959. p. 399.
551. B. Vonnegut, *J. Geophys. Res.* **65**, 203 (1960).
552. B. Vonnegut, C. B. Moore, and C. K. Harris, *J. Meteorol.* **17**, 468 (1960).
553. B. Vonnegut, *Thunderstorm Theory,* in *Problems of Atmospheric and Space Electricity,* S. C. Coroniti, ed., Elsevier Amsterdam, 1965.
554. B. Vonnegut and J. R. Weyer, *Science* **153**, 1213 (1966).
555. A. Wagner, *Meteorol. Z.* **56**, 350 (1939).
556. Waitz, *Z. Angew. Meteorol. (Wetter)* **7**, 192 (1890).
557. B. Walter, *Meteorol. Z.* **26**. 217 (1909).
558. B. Walter, *Physik. Z.* **30**, 261 (1929).
559. A. A. Ware and F. A. Haas, *Phys. Fluids* **9**, 956 (1966).
560. A. Wartmann, Sitz. Genfer Physik. Gesellsch., Dec. 20, 1888; *Arch. Sci. Phys. et Nat.* (3) **21**, 75 (1889); *Meteorol. Z.* **6**, 119 (1889); *Wetter* **6**, 90 (1889); *Bull. Soc. Astron. France* **11**, 298 (1897).
561. W. Watson, *Phil. Trans. Roy. Soc.* **48**, 765 (1754); **10**, 525 (1809 abridg.).
562. W. K. R. Watson, *Nature* **185**, 449 (1960).
563. B. E. Waye, *Nature* **155**, 752 (1945).
564. H. Webber, *J. Roy. Astron. Soc. Canada* **1**, 44 (1907).

565. L. Weber, *Meteorol. Z.* **2,** 118 (1885); *Am. Meteorol. J.* **2,** 142 (1885).
566. L. Weber, *Wetter* **13,** 167 (1896).
567. L. Weber, *Meteorol. Z.* **28,** 582 (1911).
568. L. Weber, *Meteorol. Z.* **32,** 22 (1915).
569. D. R. Wells, *Phys. Fluids* **7,** 826 (1964).
570. D. R. Wells, *Phys. Fluids* **9,** 1010 (1966).
571. W. Westphal, *Naturwiss.* **19,** 19 (1931).
572. *Wetter (Z. angew. Meteorol.)* **6,** 119 (1889).
573. *Wetter* **7,** 212 (1890).
574. *Wetter* **30,** 213 (1913).
575 *Wetter* **31,** 215 (1914).
576. J. Wilkinson, *J. Roy. Astron. Soc. Canada* **25,** 322 (1931).
577. G. Winchester, *Science* **70,** 501 (1929).
578. *Wojskowy Przeglad Lotniczy (Military Aviation Review)*, No. 12, 12 (1966); *Foreign Sci Bull. (Library of Congress)* **2,** No. 4, 52 (1966);
579. F. Wolf, *Naturwiss,* **31,** 215 (1943).
580. F. Wolf, *Naturwiss.* **43,** 415 (1956).
581. K. Wolf, *Prometheus* **26,** 229 (1915); *Meteorol. Z.* **32,** 416 (1915).
582. R. W. Wood, *Phys. Rev.* **35,** 673 (1930).
583. R. W. Wood, *Nature* **126,** 723 (1930).
584. E. R. Wooding, *Nature* **199,** 272 (1963).
585. E. J. Workman, *J. Franklin Inst.* **283,** 540 (1967).
586. R. F. Wuerker, H. Shelton, and R. V. Langmuir, *J. Appl. Phys.* **30,** 342 (1959).
587. R. F. Wuerker, H. M. Goldenberg, and R. V. Langmuir, *J. Appl. Phys.* **30,** 441 (1959).
588. A. Wüllner, *Ann. Phys., Jubelband,* 32 (1874).
589. V. V. Yankov, *Zh. eksp. teor. Fiz.* a) **36,** 560 (1959); *Soviet Phys,—JETP* **9,** 388 (1960); b) *Zh. eksp. teor. Fiz.* **37,** 224 (1959); *Soviet Phys.—JETP* **10,** 158 (1960).
590. Asim Yildiz and P. A. Silberg, *Phys. Fluids* a) **7,** 96 (1964).; b) **7,** 1721 (1964).
591. *Z. Elektrotech.* **12,** 73 (1894).
592. Ch-V. Zenger, *Compt. rend.* **109,** 294 (1889).
593. W Zschokke, *Prometheus* **14,** 234 (1903)
594. Zurcher et Margollé, *Trombes et Cyclones*, Hachette et Cie, Paris, 1876 p. 101.

Subject Classification of References

Ball lightning theories and experiments
   Chemical reaction globes (including combustion)
      32, 35, 42, 47, 100, 113, 162, 164, 208–211, 260, 280, 354, 368, 396, 428, 436, 460, 466, 480, 488, 512, 546
   Condensation of electric matter and Leyden spheres
      7, 46, 93, 309–311, 352, 363, 369, 374, 383, 388, 392, 430, 436, 495, 506
   DC electrical discharges
      110, 111, 125, 139, 147, 151, 158, 200, 205, 207, 226, 234, 237, 245, 273, 278, 288, 300, 301, 337, 340, 357, 360, 385–387, 390, 391, 398, 405, 432, 440, 458, 483, 514, 515, 517–519, 522, 527, 532, 533, 538, 539, 545, 548, 551, 557
   Dust and droplet assemblies
      7, 92, 161, 162, 233, 278, 289, 291, 331, 354, 369, 416, 417, 440, 451, 465, 500, 546, 586, 587
   Electromagnetic (ac) discharge plasmoids
      8, 9, 27, 95, 119, 167, 170, 195, 201, 208, 214, 221, 241–243, 250, 261, 277, 291, 303, 334, 404, 405, 429, 435, 475–479, 520, 525, 552, 562, 582, 586–588
   Gas dynamic structures
      73–75, 85, 109, 122, 139, 148, 149, 161, 183, 260, 265, 268, 292, 311, 322, 332, 337, 379, 465, 551, 552, 581, 584
   Globes sustained by nuclear reaction
      15, 61, 97, 121, 277
   Plasma spheres
      33, 73–75, 85–87, 118, 128, 130, 156, 183, 187, 211, 213, 236, 241, 266, 292, 293, 310, 338, 339, 356, 374, 382, 405, 430, 449, 473, 474, 476, 477, 503, 520, 529, 535, 552, 562, 581, 584
   Spherical forms from zigzag lightning
      73–75, 156, 187, 240, 252, 256, 260, 309, 310, 322, 348, 360, 361, 375, 379, 389, 503, 529, 535
   Vaporized incandescent spheres
      65, 200, 237, 293, 328, 355, 398, 477, 524, 528, 535, 538
Collections of ball lightning reports (>10)
   16, 45, 65, 101, 153, 164, 165, 181, 224, 230, 282, 321, 323, 420, 439, 452, 453, 462
Negative views on the existence of ball lightning
   31, 43, 149, 197, 223, 224, 252, 268, 272, 287, 295, 296, 306, 455, 461, 511, 541
Observations of ball lightning (1–10)
   1, 3–5, 13, 18, 19, 21–26, 28–30, 34, 38, 39, 48–54, 60, 62, 63, 67, 70–72, 76–83 (Bradyte), 84, 86, 88–90, 94, 96, 99, 103–105, 107, 112, 114, 115, 117, 124–126, 128–131, 134, 135, 137, 138, 140, 142, 143, 146, 152, 154, 155, 159, 163, 166, 168, 169, 171, 173, 175, 177–180, 182–184, 186, 188, 190, 192–194, 196–200, 202–204, 206, 209, 211, 212, 215–218, 220, 222, 224, 225, 228, 229, 232, 233, 235, 238, 239, 241, 244–246, 251, 252, 254, 256–260, 263, 265, 268–271, 274, 277, 279, 284, 286,

287, 295, 297–299, 301, 303–305, 308, 310, 316, 317–320, 324–330, 333, 335, 336, 342, 345, 346, 349, 353, 357–364, 367, 369, 370, 371, 376–378, 380, 384, 388, 393–395, 397–400, 403, 406, 407, 410, 413–415, 417–419, 422–427, 429, 431, 438, 441–448, 456, 459, 463, 464, 468, 469, 471, 472, 479, 484–488, 490–500, 502, 504, 505, 507–509, 513, 519, 522–524, 526, 528, 533, 536, 537, 542–546, 548–550, 554–556, 560, 561, 563, 564, 566–568, 571–579, 583, 591, 592, 594

Photographs
  Lightning and electric discharges
    2, 11, 38, 65, 91, 210, 219, 231, 277, 313, 361, 365, 503, 527
  Photographs of ball lightning traces
    123, 127, 227, 231, 241, 248, 277, 312, 380, 381, 401, 402, 411, 412, 418, 489, 503, 580
  Photographs of bead lightning
    42, 231, 295, 454, 457, 470
  Photographs and sketches of stationary ball lightning
    18, 108, 135, 171, 192, 233, 248, 260, 263, 272, 324, 361, 362, 450, 459, 503, 521, 542, 579
  Questionable photographs of ball lightning
    37, 127, 227, 302, 350, 351, 381, 401, 402, 411, 412, 467, 479, 501, 558, 580, 593

Plasma physics
  10, 20, 55, 56, 58, 59, 64, 98, 106, 116, 120, 132, 133, 136, 141, 172, 174, 185, 191, 201, 234, 247, 249, 262, 266, 275, 276, 283, 285, 292, 334, 373, 408, 421, 434, 437, 473–475, 525, 540, 547, 559, 569, 570, 589, 590

Properties of ball lightning
  36, 65, 108, 181, 230, 277, 315, 321, 338, 350, 372, 420, 452, 453, 465, 479, 516

Properties of lightning
  44, 66, 68, 69, 144, 164, 184, 189, 214, 221, 243, 250, 253, 255, 261, 289, 290, 294–296, 307, 343, 344, 347, 365, 366, 409, 451, 460, 461, 481, 488, 504, 514, 518, 519, 530, 534, 546, 553, 585

Reports from ancient or historical sources
  12, 17, 41, 53, 62, 72, 102, 145, 158, 160, 176, 240, 294, 352, 436, 471, 561

Reviews of the ball lightning problem
  6, 14, 16, 32, 36, 42, 45, 60, 65, 97, 101, 108, 153, 157, 164, 181, 200, 213, 230, 260, 267, 272, 277, 281, 282, 292, 315, 321, 323, 339, 350, 357, 361, 382, 420, 433–435, 449, 452, 453, 460, 465, 479, 481, 482, 488, 495, 498, 503, 510, 565, 579

Round forms observed with zigzag lightning
  38, 40, 51, 57, 68, 103, 124, 228, 230, 256, 314, 341, 353, 361, 389, 591

Index

Active nitrogen, 3, 94, 105, 114
Affixed vs free ball lightning, 63, 64, 68
Aniol, R., 77
Arago, F., 9, 18, 19, 24, 47, 48, 77, 81
Aristotle, 5, 11, 80
Association of lightning and ball lightning, 6, 9, 11, 21, 23, 25, 27, 28, 29, 32, 33, 40, 43, 46, 59, 63, 74–75, 80, 81, 94, 98, 99, 129, 146

Ball lightning
  at high altitude, 95–96
  from linear lightning, 33, 80, 94, 125, 127, 128, 129
  in airplanes, 38–42, 73, 110
  in closed rooms, 37, 42, 80, 135
  in cyclones, 42, 74, 95–96
  in tornados, 19, 43, 74, 95–96
  over water, 31
Ball-of-fire discharge, 104
Barry, J. D., vii, 50, 65, 66, 67, 72, 73, 74, 76
Bead lightning, 80, 105, 147
  from linear lightning, 24–25, 54, 110, 112, 129
  photographs, 24, 54
Benzene, 84, 86
Berger, K., 75
Bouncing ball lightning, 21, 32, 35, 64, 81, 96
Boyle, R., 38
Boys, C. V., 1, 68, 94
Bradytes, 6
Brand, W., monograph, *Der Kugelblitz*, v, vi, 50, 52, 62, 65, 77, 104
Brush discharge, 3, 10, 21, 98, 104, 113
Buoyancy of ball lightning, 106, 107, 111, 113, 136, 137, 142

C_{17} carbon atom polymer, 82
Cavendish, H., 81
Change in size or color of ball lightning, 73, 93

Charge separation in storms, 1–2, 14–15
Chemical reactions in ball lightning, 81–88
Color of ball lightning, 2, 62–64, 67, 82, 83, 85, 86, 95, 103, 112–113
Combustible fuel fireballs, 9, 68, 78, 84–86, 147
Composition of lightning channel, 14
Condensation of reactive matter in ball lightning, 78
Confinement of electrical particles, 115–120
Confusion of meteors and ball lightning, 5, 6, 18, 19
Continuing current in lightning channel, 12–13, 104, 110
Contradictory ball lightning reports, 62
Cool flames, 86
Corona discharge, 20, 44, 70–71, 96, 102
Cougnard deionizer, 71, 96
Criteria in reproduction of ball lightning, 148
Crookes, W., 101

Danger in position of ball lightning, 100, 104
Dark ring plasmoids, 142
Debye shielding distance, 115
Dmitriev, M. T., 30
Droplets, 14, 85, 90–91, 143
Duration of ball lightning, 32, 63, 67, 77, 81, 88, 93, 94, 97, 104, 107, 109–110, 113, 130, 133–134, 141, 145
Dust-devil electric field, 95
Dust particles, 78, 79, 86, 88–91

Eaborn, C., 2
Effects of ball lightning, 27–48, 73
Electric charge generation in storms, 1, 2, 14–15
Electric induction formation of ball lightning, 80
Electric field with ball lightning, 34–35

Electroluminescent air, 109–110, 140–142
Emission of sparks and rays from ball lightning, 3, 31–32, 37, 39, 43–44, 51, 58–61, 63, 72, 80, 90, 99
Energy of ball lightning, 68–69, 72, 82–83, 86, 92, 94, 96, 103, 111, 114, 131, 132–134, 143
Excitation of rf by lightning, 110, 131, 135, 143–145
Explosion of ball lightning, 7, 27, 28, 35–40, 45, 46, 59, 63, 70, 72, 73, 92, 108, 135
Externally *vs* internally powered ball lightning, 77, 93, 125, 131, 133–134, 136, 138
Externally powered ball lightning, 77, 133–145, 147

^{17}F, 88
False photographs, 52, 54–55
Faraday, M., *v*, 9, 19, 98
Finklestein, D., 107, 139, 142
Flammarion, C., 48
Focusing of current in ball lightning, 106, 108
Formation of ball lightning, 3, 6, 9, 33, 74, 78–114, 125–145, 147
from lightning, 25, 27, 75, 80, 87, 88, 98–111, 125, 127, 128, 129, 132, 134, 141–142, 147
Franklin's lightning experiment, 1
Frenkel, Y. I., 89, 96
Frequency of ball lightning, 74, 75
Fuel fireballs, 84–87
"Fulminating matter," 81, 114, 126

Galli, I., 49
Gases from ball lightning, analysis, 31–32, 83–84
"Globular vim," 114

Harris, W. Snow, 10, 48, 98, 104, 111–112
Heat from ball lightning, 67–68
Hesehus, N. A., 101–102
Hill hydrodynamic vortex, 96, 119, 127
Humphreys, W. J., 20, 21, 50
Hydrogen, 33, 83, 84, 87
Hydrogen–oxygen ratio from ball lightning, 83–84

Ionization in reversed field of lightning, 107

de Jans, C., 49, 77, 102, 104, 113
Jensen, J. C., 58–59

Kapitsa, P. L., *v*, 75, 133, 135, 145

Kelvin, Lord, 19

Laser plasmoid, 134
Leonov, R. A., 50, 77
von Lepel, F., 100
Leyden jar fireball, 9, 78–80, 100
Light
from ball lightning, 67, 109, 133–134, 147
from charged droplets, 89, 91
from lightning, 12, 13, 81
from long lived rf discharges, 109–110, 139–142
Lightning
current, 11
diameter, 13, 105
forms, 1, 5, 6, 11, 12, 15–17, 81
potential, 11
processes, 1, 2, 11–17
temperature, 13, 14, 126
Lodge, O., 104
Loeb, L. B., 21, 22
Lomonosov, M. V., 9
Long-lived fireballs, 139–142
Lucretius, 5, 11

Marsh, O. C., 38
van Marum, 112
McNally, J. R., Jr., 50
Metal globes, luminous, 33, 112–114, 142, 147
Methane, 84, 86, 87
Microwave formation of ball lightning, 131–145
Microwaves from charged droplets, 91, 143
from lightning, 144–145
Microwave sources in storms, 135, 143–145
du Moncel, Th., 98
Motion of ball lightning, 2, 3, 11, 19, 31, 32, 69, 87, 103, 105, 136
against the wind, 3, 62, 69, 105
Moving camera photographs, 52–57
Multiple lightning strokes, 12, 104, 110
Muschenbroek, P. van, 6, 9, 78

N_2 ($A^3\Sigma_u^+$), 110 141
N_2 ($w^1\Delta_u$), 141
N_{12} nitrogen atom polymer, 82
Natural electric field, 11
Nitration, 82
Nitrogen dioxide, 32, 69, 82, 102, 140, 141
Nitrogen triiodide, 81
Nitrous oxide, 82, 87, 140

Index

Nonexistence of ball lightning, 10, 18, 77
Nuclear reactions, 88
^{15}O, 88
O^-, 131
O_2^-, 131
O_2^+, 86, 142
O_2 $(a^1\Delta_g)$, 109, 141
0_2 $(b^1\Sigma_g^+)$, 109, 141
O_{12} oxygen atom polymer, 82
Odor
    of NO_2 from ball lightning, 28, 32, 46, 71, 82
    of ozone from ball lightning, 29, 32, 37, 38, 71
    of sulfur from ball lightning, 7, 29, 35, 38, 43, 71, 112
Optical illusions in sightings, 10, 19, 26, 27, 28, 46–48
Oxidation, 81, 83–87
Oxygen, 83, 84, 94, 141
Ozone, 32, 69, 82–83, 92
Ozone–nitrogen dioxide ratios from discharges, 32, 68, 83

Path of ball lightning, 2, 3, 19, 21, 25, 32, 35, 62–64, 69
    through small orifices, 70
Pentane, 86
Phosphene, 48
Pinch lightning, 25, 57, 129
Planté, G., 24, 99–101, 112
Plasma instability, 120–123, 132
Plasma sphere, v, 114, 117, 119, 122, 131
Plasmoid, 114, 115, 120, 123–125, 130, 132–133
"Ponderable matter", 114
Positive charge of ball lightning, 100, 107, 108, 131, 137, 141
Powell, J. R., vii, 107, 139, 142
Power requirement of atmospheric plasmoid, 103, 110, 113, 136–137, 139, 147
Propane, 84, 85
Properties of ball lightning, 2–3, 62–76
Protection from ball lightning, 70, 71, 73, 96

Radiation from ball lightning, 32, 128
Radioactivity of ball lightning, 88, 128
Rayle, W. D., 50, 75
Rayleigh, Ld., 105–106
Reality of meteorites, 18

Recombination, 80, 90–93, 113–114, 126, 129, 130
Richmann, G.-W., vii, 8, 9
Righi, A., 101
Ring structure in ball lightning, 118, 120, 122, 125, 126–129
Rotation of ball lightning, 36, 42, 46, 70, 94–97, 137
Round forms in linear lightning, 25–27, 80

St. Elmo's fire, 7, 19–21, 38, 40, 44, 50, 67, 71, 76, 105, 113
Sausage instability in lightning, 25, 32, 129
Sauter, F., 49, 62, 75, 77, 89
Self-confinement of plasmoids, 124–125, 127, 128, 132
Sestier, F., 48
Shape of ball lightning, 62, 67
Ship-board ball lightning, 7, 38
Silberg, P., 77
Size of ball lightning, 63, 65–66
Sound from ball lightning, 3, 32, 34, 36, 46, 57, 63, 71–72, 84, 90, 99, 101, 145
Sound from lightning, 21
Spherical discharge on photographic plate, 102, 112
Spherical flame, 85
Stekolnikov, I. S., 57
Superconducting core in ball lightning, 132
Swithenbank, J., 47

Temperature of ball lightning, 32, 68, 83, 85–86, 92, 130
Thunder and lightning, 21
Toepler, M., 101–105, 113
Torus, 67, 118, 122, 127, 128, 132
Trowbridge, J., 101

Velocity of ball lightning, 63, 70
Virial theorem, 125, 132–133
Volcanic lightning, 15, 16, 37, 42, 89
Vortexes generating ball lightning, 89, 93–97, 114, 125, 126–128, 133

Weber, L., 77
Wood, R. W., 46

p-Xylene-azo-β-naphthol, 89
Zielkiewicz, Z., 3